The Portfolio Guidebook:

Implementing Quality in an Age of Standards

The Portfolio Guidebook:

Implementing Quality in an Age of Standards

Richard Koch
and
Jean Schwartz-Petterson

with chapters by
Aaron Schippert and Alan Willoughby,
Kathryn Bell and Liz Webb

Christopher-Gordon Publishers, Inc.
Norwood, Massachusetts

Credits

Every effort has been made to contact copyright holders for permission to reproduce borrowed material where necessary. We apologize for any oversights and would be happy to rectify them in future printings.

Student work from Hillsdale used by permission of the Directory Information section of the Parent and Student Handbook of Samuel J. Gier Elementary School, granted by Alan Willoughby.

The Bill Harp Professional Teacher's Library
An Imprint of
Christopher-Gordon Publishers, Inc.
1502 Providence Highway, Suite 12
Norwood, MA 02062
(800) 934-8322

Printed in the United States of America

10 9 8 7 6 5 4 3 2 1 05 04 03 02 01 00

Library of Congress Catalogue Card Number: 99-66214
ISBN: 1-929024-01-0

Dedication

We dedicate this book to our parents
Blanche and Earl Koch, and *Genny and Ken Schwartz.*

I also dedicate this book to *Laura.*
—Richard

I also dedicate this book to *Jeff* and *Michael.*
—Jean

Acknowledgements

We are very grateful to Adrian Public Schools for their participation in the three-year portfolio research project, for their valuing of this study, and for their technical assistance.

In addition, we would like to thank Addison and Hudson Community Schools for their participation in the original study.

We are also grateful to Adrian College for its support of this collaborative research project and for its interest in helping improve public schools.

We are especially grateful to the original research team for their several years of work to develop the steps and materials upon which this book is based. That team deserves much credit for their persistence and dedication. The team included Paula Bowman, Dick Clark, Maxine Diller, Sally Kruger, Larry Roberts, and Eleanor Wollett.

We have noticed in other book acknowledgements that often one special support person is credited with saving the authors through tireless help and special knowledge of technical computer maneuvers. For us, that person is Nydia Mendez. Near the end, Nydia would arrive (called by us to come in from home) and greet us by saying, "When will your book be finished?"

We would answer, "It's just like we told you before. Today is the last day."

Finally, we are deeply indebted to Max Carter, whose time at the Lenawee Intermediate School District was marked by consistent devotion to thoughtful improvement of schools, who showed the vision to support and facilitate this work toward authentic assessment, and whose personal grace allowed so many teachers to lead.

Contents

Introduction: Why Another Book on Writing Portfolios?

❖ ❖ ❖ ❖ ❖ ❖ ❖

For some time now in the world of education there has been an unhealthy tension between workshop, or "hands on," classroom teaching practices recommended by an increasing body of research, and the usual evaluations and assessments still in place at the local, state, and national level.

Donald Graves, Nancie Atwell, Lucy Calkins, Linda Rief, Tom Romano, Carol Avery and many others have put forth clear, powerful reasons for moving to workshop approaches, as well as presenting examples and effective teacher-help guidelines for doing so. Many teachers have taken their advice and guidance and implemented those teaching practices. However, evaluation and assessment practices have largely remained as before. This means we have school grading in Language Arts often based on things like grammar and spelling tests, and state and national measures still mostly occurring in the form of standardized tests or one-shot writings.

The glaring inappropriateness of the wide use of old fashioned multiple choice tests to "measure" students' achievement in a writing-reading workshop classroom has led, however, to a small but growing innovative response in assessment practices.

Critiques of traditional practice such as *Report Card on Report Cards*, edited by Tara Azwell and Elizabeth Schmar, and the presentation of dramatically healthier alternatives to traditional evaluation such as that in *From the Child's Point of View*, by Denny Taylor, have caused some rethinking.

This type of work is joined in the current assessment conversation by an increasing number of texts by educators responsive to the classroom move to more authentic learning opportunities and workshop teaching practices. These books often explain clear, current, better practices being developed, or already in place, at lead locations, such as those at Central Park East High School or the Prospect School. Or they provide examples of practices under study at various schools, such as the increasingly wide use of the Primary Language Record originally borrowed by the United States from Great Britain. Two examples of such helpful collections would be *Expanding Student Assessment*, edited by Vito Perrone, and *Authentic Assessment in Action*, by Linda Darling-Hammond and others.

Among the alternative range of assessment practices being widely discussed and considered today, the most commonly mentioned and perhaps the one approach with the most power and promise is portfolio-based assessment.

However, as soon as a "new method" becomes widely advocated, an unfortunate side-effect of that level of recognition becomes that diminished versions of it are quickly brought into publication and into school practice. For example, it is growing more common at the state level and through publishing houses for flimsy concepts of portfolios to be presented with great fanfare. State assessment offices sometimes declare two pieces of student writing to be a "test portfolio," and in that way hinder the development of a concept of what a rich portfolio might include. Schools find themselves collecting student work and calling that collected work portfolios, often without having a clear idea of what to do with the work once it is collected.

Fortunately, there is now a large and growing number of extremely helpful portfolio books. There are books with chapters on how to make use of portfolios at different grade levels, how to make use of portfolios to assist students with revising toward quality, how to use portfolios to help students develop self-reflection abilities, and so forth. There are now so many such books that another problem, in addition to the reductive practices of some states and some schools, is that almost a blur of good books exists on portfolios.

Given this set of circumstances we're describing, it might seem that there is already enough, if not too much, help available on how to "do" writing portfolios. Certainly enough has been said to both clarify and complicate much about the issue. So, why this book?

We've been reading those portfolio materials that have been coming out over the past several years. We think a lot of them *are* good. We like the books that have now become standard reading for most folks trying to work on writing assessment, such as *Portfolios in the Reading and Writing Classroom* by Robert Tierney and his colleagues, and *Portfolio Portraits* by Donald Graves and his colleagues, as well as the book on Writing Portfolios put out by Marianne Smith of the National Writing Project, and Geoff Hewitt's view of the collected portfolio materials from the Vermont statewide assessment presented in his book, *A Portfolio Primer*. We also appreciate recent additions that diversify our sense of how portfolios can serve educators such as *Portfolios as a Learning Strategy* by Carol Porter and Janell Cleland.

And, finally, we are glad to see the signs of a stronger movement to make use of portfolios for more systematic assessments represented, for example, in "Literacy Portfolios for Teaching, Learning, and Accountability: The Bellevue Literacy Assessment Project," by Sheila Valencia and Nancy Place in *Authentic Reading Assessment*, and in "Portfolios in Large-Scale Assessment: Difficult But Not Impossible," in the Fall, 1995, issue of the journal *Educational Measurement: Issues and Practice*.

We find all of these works both conceptually helpful and practically useful. And together they provide quite a few different perspectives on portfolios. However, we haven't found any one book (nor, actually, any group of books) that provides an actual step-by-step complete system. We propose to offer here such a system, from the ground up and across all grade levels—from purposefully gathering student portfolios, reflecting meaningfully on them, and then validly and reliably rating them with statistical success that will also feed meaningfully back into the classroom.

We have completed a three-year portfolio research project ourselves where we worked through all of these issues in our own way. We began with a need to figure out what portfolios would be for us, and ended with the actual reliable rating of portfolios in such a way that we could document progress of student writers and better plan for furthering their writing development. We carefully examined and rated work of a group of early elementary students as they went through first, second, and third grades; a group of middle school students as they went through sixth, seventh, and eighth grades; and a group of high school students as they worked through tenth, eleventh, and twelfth grades.

Because we had to figure out how to do it ourselves, we can now explain how we decided what a portfolio is, how we established it as different from the writing folder in which the student's writing is kept throughout the year, what we decided to look for in the writing in order to judge growth or development, and how we made our viewing of the portfolios into a "system" that allowed us to reliably and validly measure and report on the students' development.

Because this is what many other educators at many other schools are still in the process of doing, we think our experience will be able to help others with this work. We know that the way we did all of this is just one way to examine writing portfolios. We also don't think our way is so wonderful that everyone should do as we did. In fact portfolios, and especially portfolio review, must be done according to the goals you have for writers at your school.

We are also aware that no one set of terms can serve to examine all the different genres of writing. On one level it would be silly to judge your students purely by our goals. However, on a more practical level, we think it can help you to take all of the complicated steps toward portfolio work if you have our steps written out as a rough guide to assist you along your way. Further, we have found our terms and guidelines applicable to a wide range of prose.

So, we've included in this book solutions to each problem as it arose, hoping that our experiences and guidelines will serve as a map to the territory. Once you know how we got to this territory you'll not only be able to get there too, but you'll also be better prepared to build the kind of structure for portfolio review that you want, which might be a lot or only a little like ours.

The writers of this book include members of both a first and second wave of researchers making use of the terms and analytical rating scales presented. The original team performed the systematic rating that yielded scores for groups of students at early elementary, middle school, and high school. In Chapter 1 we outline the steps we took to problem-solve, step-by-step, in order to establish our three-year initial study.

In that chapter, Dick and Jean write about how they learned to collect parts of student portfolios at several times during the year to lessen the end-of-year panic that can occur. The chapter also explains the steps the original team progressed through to train themselves as raters and then to conduct a summer rating week to reliably rate portfolios. This study examined writing gathered from three different school districts. The results from this modestly large-scale portfolio rating project are reported in the appendices of the book.

In Chapter 2 Dick uses one first grader's portfolio as a case study to write about how rating of student portfolios can more accurately reveal and precisely document growth in writing ability than perhaps any other method.

Throughout Chapter 3, Jean writes about her experience as a first grade teacher using our portfolio terms and rating scales along the way, day-to-day, to help her view student writers more clearly and to make the best possible teaching decisions about what to focus on in her classroom writing workshop.

Subsequent to that earlier work, the writers of the upper elementary chapter (Chapter 4) and the high school focused chapter (Chapter 5) joined the team. With the help of the various parties and partners, we hope we have been able to achieve several different goals and establish several clearly defined ways that our portfolio terms and rating scales might be of use.

Aaron and Alan explain in Chapter 4 how they used our portfolio system at the upper elementary level to examine students' progress and then assessed that progress as a means to reflect on their own teaching, to see what they were and were not accomplishing with their students as writers. They used the portfolio system to become more precise in noting their teaching accomplishments for a selected period of time and then to plan what to change in their writing classrooms.

Liz and Kathryn, our high school teachers, report in Chapter 5 on their work using portfolios and terms to help them teach student reflection. They guided students to use the portfolio rating system to examine their progress and to plan next steps and overall goals for themselves as writers. To do this, they had to develop an understanding of how to teach portfolio criteria carefully to students so they could come to "own" the criteria. They also needed to teach the elements necessary to make up a quality reflective statement. Liz and Kathryn were interested not only in getting good writing from students, but also in finding out how much students could take over the defining of quality and the application of criteria before the sequence of writing workshop experiences was over.

Finally, Jean, drawing on her more recent work with staff development for elementary language arts, wrote Chapter 6. She explains the approaches she took to using our system for helping other elementary teachers reflect on and improve their teaching. A key to making portfolios effective in a whole school district is having a plan to extend portfolio use through professional development. Jean describes her common sense and careful work with groups of teachers through which she addressed this need.

In Chapter 7, Dick and Jean conclude with their final recommendations and "pep talk." We know it can seem like a lot of extra work to really become systematic in using portfolios, but we give you our best suggestions and a list of reasons why we think this portfolio work is necessary to achieve the fully effective teaching that the public expects and our students deserve.

Portfolios have been around long enough for a full range of concerns about them to arise. Long enough, for example, for the arguments to be made that either portfolios are much richer than most other assessments and are underrated or they are only one way to approach assessment and are overrated. Some argue that portfolios are one more thing there is no space and time for in the sane writing classroom or, others say portfolios are necessary to achieve a sane writing classroom but are too complicated to use for large-scale assessment. These and other arguments continue today.

What we have found is that portfolios *are* necessary for a sane writing classroom. Complicated as they are, they can still be made quite straightforward, easily gathered, and fruitfully assessed, even on a moderately large scale. Portfolios can be richly rewarding for all involved in the educational enterprise: administrators, teachers, students, and parents.

Chapter 1

What is a Portfolio?
When Do You Need One?
What Goes Into It?
How do You Gather It?

Portfolios are different things to different people. And portfolios are different things depending on what you want to do with them.

Some educators feel that students should be completely in charge, choosing the works that are considered "the portfolio" and choosing the ways that they will reflect on this work. We do not quarrel with that definition, so long as the purpose is to emphasize student ownership and to see the results of student reflection over time approached in a fairly natural way.

More common as a definition is something on the other extreme, where a state government department of education will establish a quite limited view of what makes up a portfolio and a quite limited purpose. Such a portfolio might include a couple of papers from a student's three or four years of high school English class work and a couple of pieces of writing from other subject areas and have as its purpose the demonstration of the student's writing experiences and ability. This might be studied to help determine either the student's "employability" or to certify the student's readiness for a certain level of college work, or both. We do not quarrel with this definition, as long as the materials are used for limited purposes, and as long as the student's relative "opportunity to learn" is also considered. Students who have not been helped by their schooling to produce the requested materials should not be irrevocably judged in a negative way. There need to be alternatives in the preparation of such portfolios and in the judging systems.

Most common of all approaches these days is the school-based portfolio that is in itself defined quite a variety of ways. In one of its forms, the portfolio consists simply of two or three pieces of writing the teacher selects at the end of the year to send on to the next year's teacher. Or it might consist of a few pieces of writing the student and teacher jointly choose to put through a special semester-end or year-end review to help determine the student's grade. These also seem to us to be appropriate definitions and uses of portfolios.

We feel that it is legitimate and often helpful for portfolios to contain more than writing samples. Performance portfolios, or professional portfolios, have been around for some time and have often included diverse materials. Such portfolios might include photographs, art work, graphs or sketches, audio or video tapes, computer discs, three-dimensional work samples, biographical data, and letters of review or recommendation. In this book we are especially concerned with the school-based writing portfolio which might well incorporate many of the above mentioned materials.

Showcase or Growth?

One of the most commonly established purposes for portfolios is to *showcase* the best of what a student has achieved. To do this a student and/or a teacher would select the best pieces by the student, or perhaps attempt to put together a collection of pieces that demonstrated the variety and richness of the student's achievement.

The other most commonly established purpose is to demonstrate the student's *growth* in ability over time. To do this the student and/or teacher would select pieces that could show the beginning ability level of the student in their time of working together along with pieces that would represent growth in ability, especially pieces that might represent the student's final ability level of that year or work period. Then, by examining the early pieces and the later pieces comparatively, the student's growth might be observed.

We feel that one of the major strengths of our approach is that it results in a portfolio that can both showcase best work and provide samples that would allow us to document growth over time.

Whose Interests? What Purposes?

In general, we feel the key preliminary practical questions to be asked about portfolios are:

- Whose interests do we most want to serve?
- What processes do we want to take the portfolios through?

On the one hand, it is legitimate to emphasize student ownership and student learning purposes. This would suggest a high degree of student choice as to what goes into the portfolio. This should include regular time spent by the student doing self-assessment of materials in the portfolio in order to judge progress, select new projects, or establish goals for further writing development. On the other hand, it is legitimate for a teacher, school, or school district to emphasize more formal assessment purposes. This would probably necessitate that educational professionals make at least some of the key choices of materials. Teachers and administrators would also establish the central basis for review and judgment.

Still, at the beginning of their work with or toward portfolios, teachers or schools often feel a little baffled by these different choices and uses. That is why we offer here the goals and steps we chose, in the hope that the way we did it will both help and encourage more teachers and schools to use portfolio approaches to teach and assess writing.

What we did, then, was to attempt a balance of student-teacher-school purposes, as well as a balance of student and teacher review of the work with a more formal assessment on a larger scale. We feel we were successful and can, therefore, recommend such a balanced approach to others whose purposes might be well-served by this perspective.

We involved students in a selection process to pull a smaller number of pieces from the writing folder at regular intervals (selecting work from each six or nine week period) to form a preliminary portfolio. This step allowed students and teacher to filter the materials before the final review. Next we asked students and their teachers to engage in a year-end selection process that was quite simple but effective. Because we ultimately intended to emphasize a formal rating of work in the portfolios, which we did according to an analytical rating scale that we devised, we needed to limit the amount of work in the portfolio.

What We Chose to Put into the Portfolios

Part One: How we gathered work to place in the portfolios

After extensive discussion by the research team, we settled upon a maximum of five pieces of writing, including all of the parts and drafts that were available for those five different pieces of writing. We first asked the student to choose the two pieces they would most wish to have placed in the portfolio. These choices were honored to the extent that the pieces were automatically included. The student was asked to write a letter of review of the writings that included statements about what they felt were strengths, weaknesses and signs of accomplishment, as well as assessments of what would be their future goals as writers. This letter was to be placed in the portfolio and was to serve as the student's introduction to the portfolio. The students knew that they and their classmates were part of a portfolio experiment and that these, or other works of theirs, would be judged to determine the professionals' view of their progress as writers over the school year.

Next, we asked the teacher, in consultation with the student but with the teacher making the final selections, to determine from the already existing preliminary portfolio three pieces to be included in a final portfolio. These were to include what the teacher judged to be a representative early, middle, and late sample of the student's writing and accomplishments for that year. The teacher then wrote a letter of review and introduction to the portfolios from her or his classroom. It explained what the writing goals for the year had been and how those had been worked toward, as well as what she or he particularly observed in terms of strengths, weaknesses, and next steps in coaching that might be productive for the students.

The student choices of "favorites" and the teacher choices of "early, middle, and late" might well overlap. Complete overlap would mean there were only three pieces in the final portfolio that could be reviewed according to the devised scale. No overlap would mean that there were five pieces (two student-chosen and three teacher-chosen). The classroom process used for collecting both sets of choices is described by the chart in Figure 1.1 and Appendix A.

Part Two: How we moved from "collecting writing" to formal rating of portfolios

Typically, in the world of teaching and evaluating writing, a disturbing gap has occurred between a school district's evaluating of writing and usual portfolio practice. In this approach, the portfolio may be used to help determine an individual student's evaluation. The teacher might evaluate both the process participation of the student and the quality of the best few papers. The portfolios might be shown to parents in conferences, or even passed on to the next year's teacher. However, no larger review of the writing development of a grade level group is attempted, at least not one based on the portfolios.

More recently, some school systems, or even state government departments of education, have moved to the gathering and evaluating of single samples of student writing. This, of course, has also been the approach taken by the National Assessment of Educational Progress. Though there is quite a bit of research clarifying that a single sample of writing gathered in short term testing conditions is not a valid measure of ability, this gathering of a single sample at least has moved the evaluation into the arena of "looking at writing."

From the vantage point of our work, however, as educators interested in moving evaluation and assessment toward more authentic and valid results, we saw it as necessary to make fundamental use of the portfolio to gauge individual *and* group writing development. This was our central goal right from the start.

So, even though we hoped for a balance between student-teacher-classroom uses on the one hand and more formal assessment purposes on the other, we knew we were also committed in this research experiment to achieving effective use of the portfolio for somewhat broader assessment.

With this goal in mind we hoped that the student's freedom to choose two pieces to place in the portfolio and to write an introduction would serve, along with the fact that the entire portfolio was their own work, to provide a sense of student ownership. And, we hoped to help participating teachers to glean more information about their students' development without unduly interfering with any teacher's preferred classroom methods. However, we knew we might inhibit a student's sense of owning the project by our imposition of our assessment purposes, and we knew a teacher might at times alter her or his usual methods in order to accommodate the needs of the portfolio project.

We were heartened by the enthusiasm of students for their involvement in what they seemed to feel was an "important experiment" and by

Figure 1.1 Sample Teacher Checklist Steps for Gathering Portfolios.

Teacher Checklist for Gathering Student Writing Portfolios

1. Invite Dick Koch to class on the day when portfolios will be selected from the year-long student writing folders.
2. Set aside class time to do the portfolio selection.
3. Ask students to select their two pieces of writing for the portfolio and to write "Writer's Choice" clearly on the top of the first page of each piece they select.
4. Then, the teacher circulates and, possibly with student assistance, selects three pieces (including all steps of each one, any multiple drafts, or any steps from free writing to more finished writing). The teacher then labels each piece as representing "Early," "Middle," or "Late," in the student's writing development. Note: The late piece should represent the peak of the student's writing achievement, according to the teacher's best judgement. This piece should be from the last few months of the school year, but it need not be the last piece written by the student.

If some of the teacher-selected pieces are the same as the student selections, the pieces are simply labeled twice at the top to indicate their "dual role" in the portfolio.

5. The teacher and students double-check to see that all pieces are dated and arranged properly.
6. The students write a letter of introduction to their portfolios including the following ideas:
   - Why the student selected the pieces she/he chose, and what she/he likes about the writings.
   - What the student hopes a reader will notice about the different writings in the portfolio.
   - How the student feels she/he has grown or developed as a writer this year, and her/his view of strengths and weaknesses at this point.
7. The teacher writes a letter of introduction to the portfolios, which clarifies, in brief:
   - What contexts or assignments the writings come from.
   - What her/his writing goals for the students were (early, middle, and late in the year)
   - How she/he views a high and low portfolio according to the Assessment Rating Form.
8. All of these materials are gathered together in a box, and the box is labeled according to the school, teacher, and class.
9. The teacher labels and adds Affective Survey's from early and late in the year to the box.

their desire to provide their best writing as they collaborated with us. And, we were comforted by knowing that the schools participating with us by providing classrooms of students to study had done so entirely voluntarily and seemingly with a clear sense that our desire for more effective and authentic assessments mirrored their own.

To prepare to review portfolios with a formal rating process requires making a further series of decisions. The decisions may perhaps be made in different sequences, but we felt that we wanted to decide what general type of rating instrument we would use first. We made this decision about type of rating instrument even before the formal beginning of the assessment project. We knew we did not want to use holistic scoring even though it is a clear approach to rating because it provides one score per piece of writing. That did not seem to us to provide as much or as precise information as an analytical rating scale which provides several ratings per piece according to different categories.

After weeks of consensus-forming discussion in which we actually examined pieces of student writing and thought about the writing in terms of teaching goals, we settled on a set of six terms/categories. The areas we decided to review were fluency, development, organization, style/voice, revision, and editing. The first step to utilizing such terms is to determine what you mean by each one, and we did this by establishing the definitions outlined in Appendix A.

One of the early decisions that must follow, once you've decided to use an analytical scale, is what range of numbers to spread your scale over. For example, you could establish a three point range which would rate work from "1" to "3" for each category or term. The problem with a small range is that it lacks the precision or refined understandings that a broader range might define. The problem with a large range is making raters highly competent in utilizing so many different rankings.

We chose fairly quickly to use either a five point or six point scale. The benefit of a six point scale is that it forces rater choice between the higher side of the scale, "4" through "6", or the lower side, "1" through "3". There is no central or "average" number to perhaps fall back on too easily. With a five point scale there is the danger of this "safe middle" that tempts raters to use the rating of "3" more often than is entirely appropriate.

On the other hand, a five point scale involves fewer categories to learn as a rater. Also, the danger of excessive use of the "safe 3" can be somewhat avoided by being aware of it and by trying to train yourselves as raters to avoid that pitfall. Further, a five point scale can provide additional precision to definitions to either side of the "3"—tending to make a "2" mean something like "not yet" and a "4" to mean something like "solid" or "effective," not "average." In the end, because we very much wanted to achieve systematic success (rater reliability) with our scoring, and for the other reasons mentioned, we chose to work with a five point scale.

Once you have defined your terms and decided what range of a scale to use, perhaps the most difficult step comes next—deciding what these terms mean when using them to establish ratings for different groups or grade levels. This can only be done through group practice with actual student writings that can lead to consensus on what a certain rating describes for a certain grade level. The opening questions look something like this: What is a "2" in organization for a first grader?

The discussion of such questions, if it is to be enriching to the classroom and fruitful to formal assessment, must be laborious. We engaged in a series of six, three-hour meetings which allowed for practice rating and much discussion. The discussion was substantially informed by classroom teachers for the grade level in question but not dominated by any one view.

The results of such discussion prepared us for a one week period of rating that occurred just after the school year ended. The first steps of this "rating week" were devoted to squaring off against the variety of practical and conceptual problems that faced us. In particular, the first steps involved a further "practice rating" by teams of raters making a preliminary study of some actual student writings, according to the guidelines in Figure 1.2 and Appendix A.

Once such a scale was in place a systematic rating method needed to be established, such as outlined in Figure 1.3 and Appendix A.

Part Three: How we reported the results

All of this rating leads, of course, directly to a need for a system of reporting the ratings. We chose not to focus on students' final standing for the year but rather more on growth and development in different categories over the year's time. We hoped to achieve this through our rating of a representative early-in-the-year piece and a representative late-in-the-year piece according to the same rating criteria. We also chose not to focus on reporting individual student scores but rather on reporting the development of the grade level group. To achieve this group focus we developed an "Amount of Change" statistical reporting chart, which we've included in Appendix B.

Even after all of this defining, rating, and reporting, the question of "what does this mean?" is not fully answered in terms of applications for teachers and their students. For this fuller meaning we required a partly narrative and partly discursive interpretation, which is provided in Chapter 2 under the heading, "Rating Portfolios, How They Show Growth at the End of the Year." There, we will take up the more specific explanations of other parts of our process, the nature of what we feel is our success as raters of portfolios, and the significance—for teachers and students especially—of our formal rating results.

Gathering Portfolios for Review: An Example from First Grade

Similar procedures and steps for gathering portfolios can be used across grade levels. Included here are the actual steps used for first grade.

What follows now is how the initial selecting and collecting worked in Jean's classroom. We think of this microcosmic classroom glimpse as the starting point on the map to effective portfolio work we hope to offer throughout this book.

Figure 1.2 Guidelines and Steps for Establishing a Grade Level Rating Scale.

Writing Portfolio Rater Instructions

1. Select two pieces to do two separate ratings on per folder—a piece that represents the student's starting point and a piece that represents the student's fullest change from that starting point. The second rater rates the same pieces selected by the first rater.
2. Each rating sheet has student level and I.D. number on it. Rater puts her/his rating code letter on each rating sheet completed.
3. Rate 10 and compare ratings. Write preliminary guide for using the definitions of terms at the grade level in question.
4. Tag exceptional improvements, great moments, and quotable material.
5. Rate remainder and compare. Try to reconcile all differences of more than one. Create a third joint rating sheet to represent compromise.
6. Write a second version of rating sheet definitions.
7. Review definitions as a whole group.
8. A third rater rates all papers with a difference of more than one in any two ratings.
9. Write a small collection of insights, guesses, and advice for next year (insights about the papers and the rating process, advice for during the year and the next summer rating process).
10. Write a positive comment and a suggestion for the teacher.

For most students, writing folders are perfect for collecting and organizing daily work. Writing can be dated each day and placed inside the folder. Folders can then be stored in a central location to prevent any writing from becoming lost or discarded. It can also become the folder from which writing is pulled or copied for use with the writing portfolio. I use this system with most students and find it to be practical and manageable. Younger students, however, have more difficulty handling the amount of loose paper involved with this process. To keep their work organized, I find writing booklets to be a useful alternative. Booklets contain thirty to fifty pieces of paper. Unlined paper, used early in the year, can provide a record of pictures, labels, and beginning sentences. As the year progresses, lined booklets serve to accommodate a greater amount of print. Booklets are

Figure 1.3 Rating Week Schedule and Practices.

Plans for Summer Assessment Process

1. Assessing to occur over three, five-hour days (9–2).
2. At four portfolios per hour, 28 portfolios can be read and assessed in seven hours.
3. Daily Schedule:

Day 1

9–12	Practice rating sessions in whole group, refinement of rating techniques, last minute questions about the process, working toward final reliability standard.
12–1	Lunch
1–2	Begin rating process

Day 2

9–9:30	Troubleshoot any problems that came up the first day.
9:30–12	Rating
12–1	Rating

Day 3

9–12	Rating
12–1	Lunch
1–2	Write "rater explanations" of what a "1" a "3" and a "5" are at the grade level the rater is working with and for each category on the Rating Form.

dated each day and stored in a designated area. I am careful to number each booklet and am committed to saving all booklets until the end of the school year.

I think my First Grade Writing Workshop follows, more or less, what has become a national model constructed for the profession by Graves, Atwell, Calkins, and others. However, to clarify how portfolios fit in, let me review.

I begin each writing workshop with a mini-lesson. These lessons are 5 to 10 minutes in length, and address those aspects of writing that deserve attention. Early in the year, mini-lessons remain simple. At first grade they may include:

1. Modeling beginning writing concepts (moving left to right, spacing between words, writing some simple high frequency words, etc.).

2. Managing writing workshop (routines, expectations, materials, etc.).
3. Learning to hear letter sounds inside words.
4. Working as an author would work (brainstorming topics, drafting a story,

with and responding to others).

As the year progresses, lessons become more involved. First graders might be:

1. Learning to add detail and description, extending thoughts as a form of revision.
2. Changing leads and endings to create interesting and inviting stories.
3. Developing pre-writing strategies.
4. Developing independent editing skills (grammar, punctuation, spelling, etc.).

Students are encouraged to think about, or attempt to apply, the topic of the mini-lesson to their daily writing. They are also encouraged to comment during conference or sharing time when they notice the theme of the mini-lesson surface in a classmate's writing.

Writing time follows the mini-lesson and usually lasts between 15 and 20 minutes. During this time, students write or conference with others about their writing. I generally mingle throughout the room and meet with students at their seats. Since my first graders sit at tables, I believe a lot of additional learning takes place as students listen to adults working with other classmates. Some years an instructional assistant is available during writing workshop. Other years I have trained parents to assist. These extra eyes, ears, and hands provide students greater access to an adult, keeping writing time focused and on task.

When writing time is finished, students bring their folders to the meeting area. There, two or three students are chosen to sit in the "author's chair" and read their writing. Classmate's respond to the piece in one of three ways:

1. What I like or remember about the piece.
2. What I notice is unique about the piece.
3. Something I still wonder about the piece.

I frequently record these comments on a "sticky note" and attach it to the piece for use the following day. I am careful to see that all students share their writing and respond to another student's writing on a regular basis. A writing workshop notebook contains names and checklists to help me organize conferencing, sharing, and publishing information.

Collecting the Portfolio

About every two months, I set aside class time to choose a selection for the portfolio. These pieces are taken or copied from the student's writing booklet and represent what is considered the student's "best work" for the previous eight weeks. Students participate in the process by selecting their favorite piece and placing the words "Writer's Choice" at the top of the page. When I agree with the student's selection, the piece is labeled twice to indicate its dual role. If I feel another piece is more suitable to the portfolio, that piece is labeled "Teacher's Choice" and is placed in the folder as well. Each selection contains any drafts, revisions, and editings that are available, as well as any conference notes or checklists that may seem beneficial. Students are also asked to reflect upon their writing and search out any areas of improvement at this time. Questions for this written reflection include:

1. Why did you select this piece for your portfolio?
2. What do you want the teacher to notice about this piece?
3. How do you feel you are developing as a writer?
4. What do you view as your strengths and weaknesses to this point?
5. What goals do you have for your writing in the coming months?

I find it a manageable task to help students select a piece and reflect upon growth every eight weeks. This provides the portfolio with four pieces of writing from writing workshop, and gives students a regular opportunity to examine personal goals and accomplishments. It also ensures that enough writing is available to the portfolio for an adequate year-end evaluation.

Making any portfolio a true indicator of student progress demands the total commitment of the teacher and the school to its discipline. Work must be dated and collected systematically, and students must be encouraged to take pride in the efforts, treasuring the portfolio as an important measure of yearly progress. Time for teachers or rating teams to adequately assess portfolios must also be arranged. Only then will thoughtful analysis occur and the goal of quality planning for each student be realized.

Even though much good work has been done nationally in the name of portfolios, there is still much to be worked out on the question of how best to gather and use them. There are still many places that are just beginning on their own portfolio journeys. It is to this unfolding conversation, and for the purpose of helping establish a strong foundation for further portfolio work, that we offer our guidebook help.

Chapter 2

Rating Portfolios and How They Show Growth at the End of the Year—Case Study of One Student

❖ ❖ ❖ ❖ ❖ ❖ ❖

Background

As we acknowledged in our introduction, portfolios are no longer a new idea. Much good help has been published about using portfolios in the last few years. This extensive work looking into portfolios as a basis for important educational assessments is a very good thing for a variety of reasons. Until now one of the biggest roadblocks to teachers using workshop methods in language arts has been inappropriate evaluations. These inappropriate evaluations take the form of standardized tests, one-shot writing tests, and traditional classroom grading practices.

We believe that integrated language arts practices, or workshop approaches, can solidly outdistance classroom drill and worksheet practice. Still, to the extent that poorly implemented workshop approaches might yield unsatisfactory results, we also believe that assessments should note this and help correct it. The teacher help literature is rich with examples of effective workshop teaching and with methods for how to achieve that. So, where there might be insufficient attention to the whole writing process, or what might be called "half-baked Whole Language," we educators should be as eager as any public to seek it out and change it.

However, just as there is a gap between concepts of best practice and some actual current practices in schools, there is also a gap between the concern often expressed about the need for quality assessment and the often disorganized portfolio practices. Conversation is still needed to fully describe how we can best make use of portfolios to make ourselves more diagnostic teachers and how a portfolio can be gathered and examined systematically to document student growth.

In hopes of contributing helpfully to that conversation, we examine here the ways in which "portfolio rating," as we practiced it, can increase the effectiveness of teacher reflection and can result in systematic evaluation of student work that still supports the integrated language arts approach.

As we proceed, we have two related but distinct purposes. First, since the actual systematic rating of student portfolios is still a relatively new possibility to many educators, we want to show how we did it. To demonstrate that part, we pick up where we left off in Chapter 1 and "walk through" the steps needed for application of our terms and definitions. This can serve as a guide for the way into this type of study. Second, we use writing from one student's portfolio as a case study to suggest how our view of writing progress can be enhanced through this systematic and analytical approach to portfolio review.

Looking at One Student

There is probably little that is more dramatic in educational observation than examining the writing progress of a child going through first grade with supportive writing workshop opportunities. Normally, we get to see the child begin by communicating through pictures and then develop into a pretty solid user of language on the page. We get to see first graders "break through" and become writers.

Such was the case with our research team as we viewed first grade portfolios. Here we examine the work of one student, Michelle, whom we view in her first grade year working with her teacher, Jean Schwartz, at Garfield Elementary School in Adrian, Michigan.

Figure 2.1 is Michelle's first week writing workshop example, a picture of a person of indeterminate age and gender accompanied by what appear to be two dogs.

As Lucy Calkins aptly points out, pictures often suggest whole stories to young children. Usually, the child has language to offer if an interested adult asks, "Would you tell me about your picture?"

By the next week Michelle has decided to add letters to her picture, which in this case is a building (probably a house) with a door and two windows as shown in Figure 2.2.

Michelle was given the suggestion that she might draw a line on her page, above which a picture could be drawn and below which some words could be added.

It is not at all unheard of, but it probably is somewhat advanced, for a first grader to accept this invitation to add language so close to the outset of the school year. We're pretty sure we're able to translate the intent of her letters accurately. Perhaps you have already done so yourself. Her language reads, "I have two brothers. His name is Michael and Mark. I have a Mom. Her name is Jill."

Presumably, in her story, the family lives in a house represented by the one in the picture. This means, for one thing, that Michelle is not just using language to explain her illustration, but she is already using the picture as a reminder of the story (or information) that she wants to tell. It is her visual anchor and signal of her "topic," but it is not something she feels obligated to explain.

Figure 2.1 First week writing workshop example.

Figure 2.2 Second week writing workshop example.

Dan Kirby and Tom Liner present this as a technique for helping high school students think of topics. They recommend what they call the "Life Map" in which the writer draws a series of quick, crudely sketched pictures representing, in more or less chronological sequence, some of the major places, moments, and relationships of the writer's life. These pictures then serve as visual reminders of the stories the writer has to tell or, of the topics she or he has to tell about.

By October (10/11) Michelle has already made a much larger commitment to the words on her paper as shown in Figure 2.3.

Figure 2.3 Second month writing workshop example.

There are 26 words in this piece compared to 18 in her September writing, almost one-third more. She was allowed to work on both over several days, until she felt finished.

The words on the bottom of the page are shaded by the picture that was drawn over them. The picture now has become an illustration of an already told story, or perhaps the stamp on the page that makes Michelle feel as if her story is completed. In her mind, at this point, writing comes with a picture. That's how it is.

Michelle is also moving rapidly to get control of language. Her spelling has progressed in about a month from vaguely phonetic to fairly close to the mark. The "hard" words that are spelled correctly are often words she has steady access to in her reading, or they may be available for copying off of some poster in the room (words like "school," "friend," and "teacher").

She has, it appears, recently been taught the mini-lesson on periods, and she is having a nice time placing periods in her writing. At the moment they seem to be a device to help her put a space between words, a new step for her.

You probably don't need this writing translated. But, for convenience, we will. Placed mostly in adult level correctness, her message reads, "I like my school. It is fun. I have two friends. Her name is Julie and Katie. I have a teacher. Her name is Mrs. Schwartz." The picture seems to be a school-like building.

Figure 2.4 is Michelle's writing in January (1/17).

Figure 2.4 Mid-year writing workshop example.

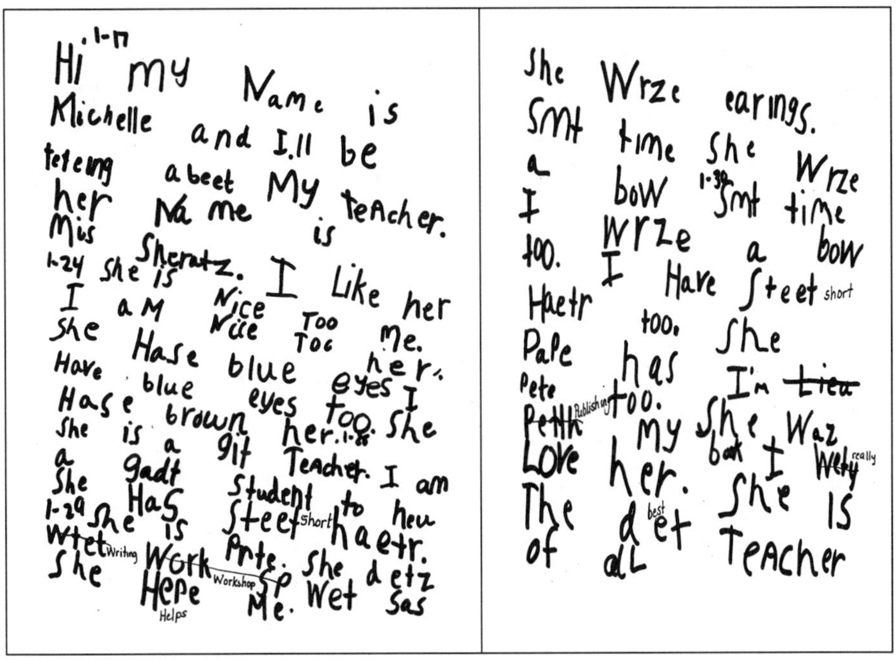

Translated into adult spelling:

Hi my Name is Michelle and I'll be telling about my teacher. Her Name is Mrs. Schwartz. I like her she is nice to me. I am nice to her. She has blue eyes I have blue eyes too. She has brown hair. She is a great teacher. I am a good student to her. She has straight short hair. She is pretty. She does writing workshop with us. She helps me. She wears earrings. Sometimes she wears a bow. Sometimes I wear a bow too. I have straight short hair too. She probably thinks I'm pretty too. She was publishing my book I wrote (I) love her. She is the best teacher of all.

Besides writing more correctly, Michelle has apparently had a mini-lesson on "beginnings" and "endings." She begins with a greeting and ends with a summary statement for emphasis. She is getting a sense of the writer's ability to craft the shape of a piece of writing.

By May, Michelle has become a writer in full bloom. Her year-end portfolio includes writings up to six pages long. The May (5/13) writing, Figure 2.5, included here was developed over several days and extends for a bit over three pages of lined paper with handwriting pretty much the same size as when she began the year.

Figure 2.5 May Writing Workshop Example.

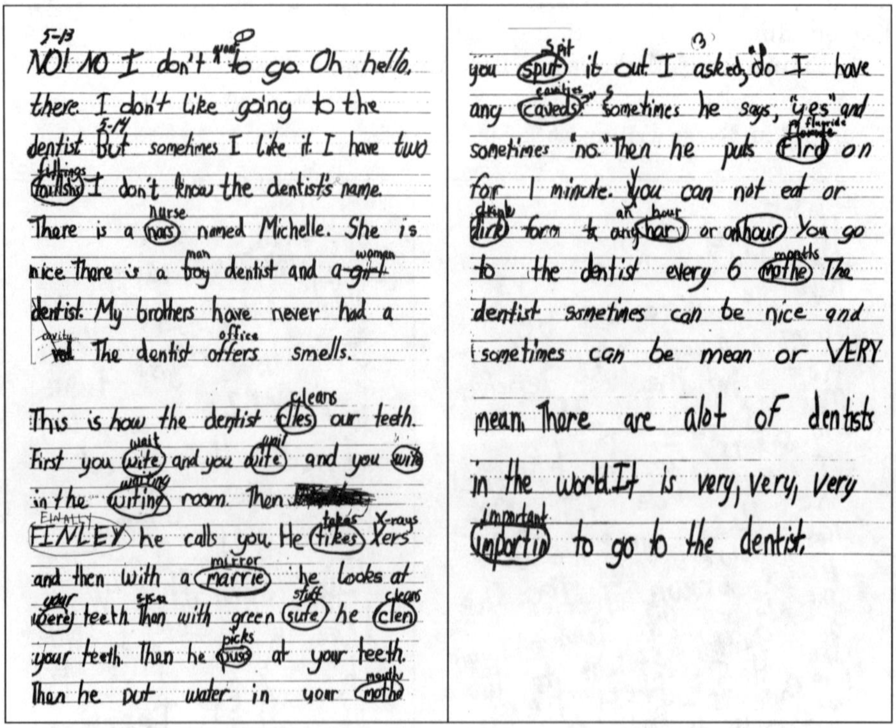

Placed in mostly adult-level correctness, it reads:

> No! No I don't want to go. Oh hello there. I don't like going to the dentist But sometimes I like it. I have two fillings I don't know the dentist's name. There is a nurse named Michelle. She is nice. There is a man dentist and a woman dentist. My brothers have never had a cavity. The dentist office smells. This is how the dentist cleans your teeth. First you wait and you wait and you wait in the waiting room. Then FINALLY he calls you. He takes x-rays and then with a mirror he looks at your teeth. Then with green stuff he cleans your teeth. Then he picks at your teeth. Then he puts water in your mouth you spit it out. I asked, "do I have any cavities?" Sometimes he says, "yes" and sometimes "no." Then he puts fluoride on for 1 minute. You can not eat or drink for one-half an hour or an hour you go to the dentist every 6 months The dentist sometimes can be nice and sometimes can be mean or VERY mean. There are a lot of dentists in the world. It is very, very, very important to go to the dentist.

Michelle is now placing spaces between words as a regular convention of her writing, and she is using periods for the correct function, to separate sentences. Her first 24 words are all spelled correctly. Throughout the piece, most of her words are spelled correctly. She once again tries for a "reader grabbing" beginning, opening with a dramatic quote, "No! No! I don't want to go." And again she ends with a sense of closure, emphasizing how important it is to go to the dentist. In between she has developed her piece with quite a bit of information, successfully keeping herself organized so that she tells it chronologically. It is clear Michelle is now writing easily and ably.

The ultimate point here is not so much a basic proof that first graders can develop well as writers. Writing workshop teachers nationwide are already aware of this. We offer this preliminary view of Michelle's progress to suggest the great value in the very existence of the portfolio—it allows us to review Michelle's growth over the year in important ways. We offer this overview of her writing in order to lay the groundwork to show what more can be seen if we take the next step to even more focused portfolio analysis.

Our Terms and Definitions: A Closer Look at Michelle

After a couple of extensive discussions, we were able, out of all the possible things teachers might look for in student writing, to decide to examine the qualities represented by the terms and definitions in Figure 2.6 and Appendix A.

Any terms and definitions that a school or group might choose will only represent a particular construct, or paradigm, through which to view student writing. Nevertheless, these are the terms that expressed some of our values for good writing, and we felt we could define them in such a way that they would serve to guide us in looking at student work.

The most basic question to be asked about such terms, after the preliminary question, "Are they clear?", is whether the construct generated by the terms is reductive or enriching as it affects our view of student work. The rest of this chapter is intended to show how our terms enriched our view.

According to the terms and definitions, our comments about Michelle's progress so far have been mostly limited to two areas: *Fluency* and *Editing*. We have been giving a little specificity to the nature of Michelle's growing ease and facility (fluency) with writing. And, we have been describing some facets of her impressive and rapid growth in correctness. To us it is startling to see how close to adult correctness she is at the end of first grade.

If we look at Michelle's final piece of writing through our other terms, beginning with *Development* (letting these brief comments on fluency suffice in that area for the moment), we will then be approximating what our portfolio readers did in their attempt to become more precise about Michelle's work.

Michelle's early pictures were not very detailed or filled out. Her early language steps also contained a pretty sparse amount of information. In

Figure 2.6 Terms and Definitions.

Writing: Terms and Definitions

Fluency

Fluency refers to "ease with which the writer writes when presented with an occasion for writing."

- So, fluency has to do purely with "amount" of writing in a given period of time, or in a given situation. We would like students to be able to generate and sustain writing more and more readily throughout their school career.
- Therefore, given a certain amount of time for journal writing, or free writing, over a period of weeks or months, we wish for the student's typical journal-write or free-write to grow longer and longer.
- Ultimately, we have no desire to encourage teachers to press a student past a "peak fluency" point, nor do we wish, overall, to press so hard for fluency, or to value it so much, that we drive students regularly toward nonsense as opposed to "thinking through writing." However, consistent fluency is one mark of a good writer who is in control of his or her writing process.

Development

Development refers to the amount of helpful information included in the piece of writing.

- Development has to do not just with number of words but with the amount of specifics (facts and/or sensory details) included. And, unlike fluency, development also has a clearly "qualitative" aspect to it.
- The amount that is desirable will vary for different genres and situations.
- Although volume is an issue, "appropriateness" to genre and rhetorical situation also are issues relevant to successful development.
- There are three basic steps in quality:
    1. The lowest quality that is still somewhat successful (at least as a practice step) is the inclusion of some facts and/or details, or an increasing number of these, but with the problem being that not all of these facts and details are related to the subject of the piece of writing.
    2. The facts and details are, for the most part, "related" to the subject of the paper, but some important information is left out.
    3. Nearly all of the facts and details are not only "related" but either "relevant" or "essential" to the subject and purpose of the piece of writing. Nothing important is left out.
- So, both volume and relevance are important aspects of development. However, volume is a valuable practice step that will later allow for the teaching of the issue of relevance.

Organization

Organization refers to the "order" of ingredients in the piece of writing and to the maintenance or focus on a central issue and related issues.

- Ordinarily, one main issue should be present in a piece, and the order of elements should clarify and enhance our understanding of that issue.
- Some obvious signs of organizational control, or lack or it, would be the way the writer handles an attempt to deliver chronology, and whether, in an explanatory or persuasive piece, the introductory material comes early and the concluding material comes late.
  - — We do not want to encourage any additional emphasis in the schools on "lockstep" organizational patterns ("always start with a thesis sentence," etc.), however, we want to assess whether parts are in a helpful or hindering order. In other words, we would assess whether clarity and meaning are enhanced or harmed by the order of parts of the piece.
  - — Organization can be an issue at the level of the whole piece and at the level of the paragraph.

Style/Voice

This category refers to the use of voice (personality) and word choice in the writing.

It also refers to the quality of control of sentence structure and sentence variety (but not correctness of sentences).

We look at this area of the writing first in terms of whether those issues are handled appropriately, then whether they are handled effectively, and finally whether they (sentence voice, and word choice) are handled with special imagination or power.

Variety

Variety refers simply to whether the writer demonstrates ability to deal successfully with different writing situations and topics. This is less essential and less reliably measured than the three categories listed above.

- If a writer is successful in the areas of fluency, development, and organization over time, that writer will likely have demonstrated success with "variety" also.
- Over time, we might wish for the upper level student to be presented a variety of opportunities for writing.
- However, since it is probably appropriate for primary students to be involved most extensively with free writing, journal writing, and learning log and observation writing, we do not want to think that these students have "failed" to demonstrate variety.

- Furthermore, apparently obsessive treatment of a particular topic by a writer is not necessarily harmful (even famous writers sometimes deal with a topic obsessively). So, we want to notice variety (teacher-offered or self-selected) over time, but we want to be very discriminating and restrained in our observation of variety.

Revision

Revision issues also will be different for different age students. In general, revision means returning to a previous draft or free writing and making it better in terms of the above areas of consideration on our rating form.

- At the primary level we are looking for the beginning inclinations in the writer to make use of feedback from others to reconsider their piece and make changes in it, and for a student's repertoire of revision strategies to gradually increase.
- We look for these developments keeping in mind the tendency at certain ages to feel "all done" as soon as a preliminary draft is completed (we do not expect a teacher to spend an excessive amount of time pressing for revisions, when it would probably be more productive for the student to be encouraged to work toward fluency and development).
- We also keep in mind that not all pieces (and in some situations, only a small number of pieces) should be revised.
- At the upper level we also look for signs of use of feedback and for an expanding repertoire of revision strategies.
- In addition, we can look for, and hope for, an increased precision and appropriateness of revisions—what is revised is the part, or aspect, that most needs revising. And, after one or more attempts, the revision is successful.

Editing

Editing refers to "correctness" issues. As indicated on the Assessment Form itself, these issues are fundamentally different from primary level to upper level.

- The degree of emphasis on this issue should be modest through elementary school and not excessive at the upper level.
- We look for correctness to improve gradually in student writing, and we look for signs of appropriate teacher help with correctness issues. Appropriate help would occur in the context of writing and through mini-lesson or workshop steps, with correctness concerns applied at the final, or "editing" stage of writing.

her first writing with words on September 4th, she did little more than name three members of her family. In her October writing she also labels the roles of three people (two friends and one teacher) and names them. However, she adds a statement of feeling, "I like my school," and she adds a brief clarifying sentence providing a reason why, "It is fun."

We can see that Michelle is not only growing in fluency but learning how to develop a piece of writing with relevant ingredients.

The piece about her visit to the dentist, written near the end of the school year, has substantially more of these ingredients (feelings and clarifications) as well as more specific information. It includes the fact that there is a nurse (hygienist?) there, that dentists can be either men or women, and the steps you go through once you are being treated by the dentist.

She is much more fully developing her writing, and she also has a broader repertoire of strategies for developing it. Because we know this classroom, we know that each daily writing workshop period normally ends with a chance for some writers to read aloud to the class, after which the class members tell what they like or remember about the piece and then ask a few questions.

Michelle participated in this process and, in response to the early parts of her piece, was probably asked questions such as, "What was the dentist's name?" (which is why she tells us she doesn't know), and "What did the dentist do to you?" and, "Did you have any cavities?"

In relation to our "three steps" toward highest quality on development, Michelle has arrived at level two. She is not just including facts and details, but she has already grasped the rhetorical concept that there is typically one main topic per piece of writing, and the facts and details should be related to that topic.

To reach the third level she would have had to arrive at the perhaps unrealistically sophisticated point (for a first grader) where she was able to recognize a mature writer's usual practice of developing key points the most. To us, her key point is yet to emerge fully. She may be intending to communicate that the visit is somewhat unpleasant and scary, but that it is important to go to the dentist anyway. She doesn't develop any information about what "mean" things might happen to you at the dentist, nor does she explain why it is "very, very, very" important to go.

This is not really to suggest that Michelle should have done this, nor that her teacher should have addressed this technical issue. In fact, Michelle is probably somewhat advanced in this area of development for a first grader. The purpose of these analytical comments is to enrich our view of Michelle's progress (and her teacher's accomplishments). This in turn helps us to clarify a possible next goal for Michelle (in second grade, perhaps)—to develop her pieces even more fully at times and with even more of the essential information needed for a reader to understand her key point.

Our definition of *Organization* is less precise than development, and may be necessarily so. The techniques and "shoulds" of development are, we think, easier to summarize. The more precise you get about organization, on the other hand, the greater the danger that you will be herding writers toward some "lock-step" pattern, something it was important to us to avoid.

However, using our definition, two things become clear about Michelle and organization. First, by the end of the year she has some consciousness of this rhetorical issue. She starts with a greeting and an orientation for the reader to the subject of the paper and then moves on to some background information. She is also aware that an ending should be special and conclusive.

Second, we can see she has not yet come to grips with the question of how adding new information may affect overall organization. It may be that the information is not most effectively included by adding to the end of what has already been written.

For example, she just adds clarifications about not knowing the dentist's name and about the issue of cavities later. The readers have asked for the name, and now she has told them she doesn't know. "Enough said," is apparently her feeling. However, this tendency to add on instead of reorganizing is something first grade writing workshop teachers will recognize as what first graders do.

Here again, the term "organization" and definition statements can help us notice her ability to start with an introductory sentence and close with a concluding idea. We are also helped to note her fairly good control of chronological order in her story. We see through our definitions with greater precision how impressively more aware of organization Michelle is by the end of the year, and what her next steps will probably be—what steps her second grade teacher might be prepared to facilitate.

We found it difficult to judge first graders with regard to *Style / Voice* issues in relation to our definition. However, two things are also quite clear with reference to Michelle's writing and her work with style and voice. At the end of the year, of course, Michelle is writing in sentences, which is quite a step. Second, Michelle's word choice is adding precision to her writing. She adds a color word for the cleaning "stuff" the dentist uses, and offers fairly specific explanations such as the fact that he "picks at" your teeth. She also includes a brief dialogue about cavities. If we stop to think about sentences and words for a moment, we are impressed at her progress.

On the other hand, her sentences are "subject, verb, completion part" sentences. At some point, probably not too far down the road in her writing life, she will be ready for the mini-lesson that might show how some of her favorite children's book writers vary their sentence patterns. Again, by being clear about what area of writing we want to examine, and by selecting representative samples from her portfolio, we see more clearly her achievement and her direction as a writer.

The readers of first grade writing in our study decided not to make use of the category of *Revision.* This was due to difficulty in deciding whether the day-by-day adding-on constituted revision and due to information from the classroom teachers that revision was not a specific goal (beyond the practice of adding on). We do not want to find ourselves assessing areas that are not appropriate to age level of the student or are not a part of the teacher's goals for that class.

We have already commented on *Editing,* or correctness, as an area of substantial progress for Michelle. It is clear that Michelle has had the opportunity to learn about how language is a different way from pictures to tell your story. She has also learned how sentences work (a sentence is a cluster of words that begins with a capital letter, ends with a period, and rides on the back of its main subject and verb). Indenting for new paragraphs was a step taught by the teachers in our study, but Michelle does not appear to be experimenting with that yet.

It is also clear that her teacher is teaching spelling in the context of Michelle's writing by adding omitted words and by correctly spelling words for Michelle as a final step of conferencing.

Because upper level criteria are described by our terms and definitions we can use them to see more clearly where Michelle might go next as a writer and what she might benefit from being coached about as she continues to write.

We are able, with these terms and definitions, to more fully describe her achievement, and in a way that will be more clear and satisfying for all stakeholders in Michelle's education: community, school administrators, other teachers, her parents, and Michelle herself.

Working From an Analytical Scale

The progress of a group, as well as that of an individual, can be examined with good precision if we further refine definitions of terms so they allow for actual rating of papers. In Chapter 1 we explained how we took the three steps necessary to enable us to do ratings of portfolio work.

1. We collected student writing in a folder throughout the year.
2. We came up with a plan for selecting representative early and late writings for each student.
3. We established rating values by grade levels, which raters then practiced using sufficiently to become reliable.

The third step, moving from more generic terms and definitions to a rating values sheet for a grade level, must involve use of the student work. Having selected from the portfolio the representative early and late in the year pieces, a team of two raters prepare to rate these pieces with the rough or generic guidance of the overall five point scale outlined in Figure 2.7 and Appendix A.

To determine what the five-category Analytical Rating Scale means for a certain grade level, the raters must read some papers from that grade level and then establish "rating values." In this work session environment for the already trained raters we found we could productively ask and answer questions such as, "What is a 'five' in fluency for a first grader?" The "First Grade Rating Values" arrived at and subsequently applied to Michelle's work and that of her classmates appear in Figure 2.8 and Appendix A.

Using something like Michelle's September 4, Figure 2.2, piece as an early writing, and something like her May 13, Figure 2.5, piece on her visit to the dentist as the late writing, the raters established early and late ratings for Michelle (who was Number 23 out of the randomly numbered 51 first graders in our study) as seen on page 28.

Figure 2.7 Writing Portfolio Analytical Rating Scale.

Writing Portfolio Assessment Form					
(effectiveness) (amount used)	**ineffective not used 1**	**low little 2**	**moderate moderate 3**	**good often 4**	**excellent always 5**
Fluency					
Development					
Organization					
Style/Voice					
Revision					
Editing (Mechanics: spelling, capitalization, punctuation, word usage; Elementary: picture, lines and letters, invented spelling, correct spelling)					

Figure 2.8 First Grade Rating Values.

First Grade Rating Values

Name ______________________ Fall ❑ Winter ❑ Spring ❑

Fluency

1. A picture
2. Signs of print awareness (copying a word or attempting to write a thought)
3. Three or more thoughts
4. Up to a page of thoughts
5. Extended pages

Development

1. A message through picture
2. One piece of information or thought (a label)
3. Three pieces of information related to a topic
4. Provides more than three pieces of information related to the topic
5. Exceptional use of facts and details

Organization

1. Message is through picture
2. Organization mostly through labeling
3. Little sense of story, limited to sequence or statement of a few facts
4. Opening sentence, details and a conclusion
5. Same as above but with definite signs that the writer consciously attempted organization

Style/Voice

1. Style and voice through picture
2. Basic information, little description, minimal signs of personality, fragments
3. Attempts sentences consistently and uses some descriptive language to show personal involvement
4. Variety in word choice; writes in sentences using descriptive language appropriate for topic
5. Includes criteria above, but with greater sophistication

Editing/Correctness

1. No evidence of correct spelling, punctuation, capitalization
2. Attempts correct spelling, punctuation, capitalization
3. Evidence of internalized sense of correct spelling, punctuation, and capitalization appropriate to age level

Student # 23

	Fluency	Development	Organization	Style/Voice	Editing
Early	2/2	2/2	2/2	2/2	1/1
Late	5/5	5/5	4/4	4/4	4/4

A quick review of these ratings can show the precision that can be achieved in viewing writing in the portfolio.

In fluency, the rating team was clarifying that "early" on Michelle was "showing signs of print awareness, copying (or attempting) to use words, and attempting to write a complete thought." By "late" in the year she was writing "extended pages" (more than a page). The fact that Michelle's longest piece was six pages shows she actually exceeded the criteria for a "5" by a significant margin.

In regard to development, the raters observed that Michelle's early work fit into the category of "labeling, including one (or two) pieces of information or thoughts," whereas in her late piece she exhibited "exceptional use of facts and details." In this category her work also exceeded the "4" rating value of "provides more than three pieces of information consistently related to topic." Hence, the work received a "5."

In organization, Michelle's early piece was represented by "organization mostly through labeling," whereas in her late piece her work was characterized by "opening sentence, details, and conclusion," including "logical sequence," rating a "4" on our scale. The raters didn't feel there was enough evidence of a "consciously attempted organization" to warrant the next higher rating.

In relation to style/voice, Michelle's early piece showed "basic information, little description, and not much sign of the writer's personality." By the late work, the raters saw "variety in word choice, writes in sentences, uses descriptive language appropriate to topic." This again exacted a "4" rating.

Finally, in editing, Michelle's early work showed her to be just where you'd expect a beginning of the year first grader to be, showing "no evidence of correct use of spelling, punctuation, and capitalization." By the end of the year raters saw her as headed toward, but not quite up to, "evidences internalized sense of correct spelling, punctuation, and capitalization at an age appropriate level." So, they awarded another "4."

The most appropriate overview statement at this point might be, "All is well with Michelle." She has come a long way as a writer in first grade and is already able to communicate effectively and with feeling. Her writing shows mostly correct usage and mechanics. What our portfolio review has provided is a much more precise and differentiated view of her progress (to better "showcase" it). The rating review of an early piece also more fully clarified where she began as a writer (to better show "growth"). Together, the early and late ratings draw a clearer map of where she might go next to improve as a writer. Our modestly large-scale review of classroom groups of students allows us to see a grade level's, or potentially even a school district's, progress in writing in this precise and convincing way.

There are cautions and precautions, however, that should be observed in taking this step of rating writing. A frequent mistake made by educators working in good faith to become precise in their analysis of student writing is to turn ratings automatically into responses. This can be very damaging. It would have done Michelle no good and probably some harm (to her self-esteem) if, early in the year, she were to receive a collection of scores that were low rating numbers for her pictures (due to the absence of words). This would be an unnecessarily negative way to document for her that she was at the beginning of her writing career.

Rather, the supportive responses her teachers and classmates did provide, along with the technical advice and editing help, were just what Michelle needed to grow as a writer.

Notifying students about comparative results among them is something that should be done only occasionally as this too will tend to replace more focused coaching that the teacher should be doing and will disrupt the focus of the writers from their topics and purposes. Finally, great care must be taken with any comparison among schools or across districts because different goals and practices would naturally lead to different, not necessarily inferior, results.

The portfolio terms, first grade rating values, and actual ratings help us see student work with a sharpness and clarity that greatly encourages and enhances a diagnostic approach to writing. One of the gifts of the Whole Language movement has been a much greater focus on teaching diagnostically and in the context of the student working on a "real" task like reading a book of choice or writing with a sincere desire to say something to someone.

Powerful and worthwhile goal setting for students and teachers can only emerge from a careful review of student achievement in relation to precise criteria for quality. The portfolio rating of Michelle's writing growth provides just such a close review and thereby a springboard to a skillfully refined and realistic sense of what might be her best next steps.

Postscript

Generally, in conversations among educators or between educators and policymakers, a clear background assumption is that regardless of what we decide to do, assessments are "human-being-neutral." This is not true. The dunking and drowning of women in American colonial times to assess whether they were witches, and the methods of torture used during the period of the Spanish Inquisition to assess whether a person was a true Christian were not neutral. Of course, these are the utmost extreme examples. However, the underlying truth importantly carries over—no assessment is neutral.

Assessments either promote and support healthy development or they interfere with it. The current high stakes state tests and the pressured one-shot writing exams pretend to be drawing a picture of the student. But this picture leaves off the hands that could have been used to accomplish authentic tasks, and it leaves out the face that could have shown us the student's personal needs and possibilities.

Workshop practices and authentic assessments are both needed by the students in our care. These two companion methods take turns delivering and receiving positive momentum from one another in a harmonious and balanced way. They take turns providing the energy needed to light the teacher's path in the classroom and to help the student to repeatedly and powerfully surge forward with learning.

Michelle no doubt would have survived in nearly any first grade classroom. She would have succeeded in most of them. However, she flourished in this case because of skilled workshop teaching. Further, it is the rich authentic assessment we have in place that is needed to help ensure that she will continue to thrive.

CHAPTER 3

Using Portfolios Diagnostically and How They Help With Teaching Decisions Day-by-Day: First Grade

❖ ❖ ❖ ❖ ❖ ❖ ❖

Jean's First Grade Classroom

I had been committed to the teaching of Language Arts skills through Reading and Writing Workshop for several years. I felt that students greatly benefited from this process and wished that our state and local assessments were more helpful in noticing these results. When I moved from third grade to first grade I continued to have these same assessment concerns.

Working with emerging writers is a challenging and stimulating experience. My students came to school with a thirst for literacy and a confidence in their ability to use language easily. They were ready to explore the world of print, and welcomed every opportunity to apply their knowledge of reading and writing to "real-life" situations. Never before had I encountered a group so eager to learn. With this kind of fervor from my students, my role as teacher seemed clear; embrace this enthusiasm with a learning experience that would honor their lived experience, cultivate natural curiosity, and foster continued growth in a caring and nurturing environment. But how was this best accomplished? How would I introduce the conventions of writing to those students who had little understanding of print? How would I further develop the talents of those who had deeper understandings? And, maybe most of all, how would I monitor the growth of each student to plan for individual need? What I needed was a procedure for documenting progress that becomes obvious only when writing is observed over an extended period of time. The assessment tool outlined in this guidebook provided me that opportunity.

Using the Rating System to Become More Reflective in My Teaching: Background

My prior experience with the writing process gave me every confidence that I could expect the same positive results with younger writers. My first step was to make sure that each student had a number of daily writing opportunities to practice and improve their skill. So each morning my first graders participated in writing workshop. They were given time to write about what was important to them, encouraging responses from adults and peers, and guidance in learning how to develop and organize their thoughts. Mini-lessons covered the entire spectrum from hearing sounds in words to developing personal style and voice. Conferences gave focus and support to the individual writer, and sharing time allowed young authors to proudly showcase their work. Students were also invited to use writing throughout the school day. Response journals, learning logs, diagrams, newspapers, lists, poetry, and observations were all part of a learning experience that had reading and writing at its core. By the end of the year the results seemed clear. Given daily opportunities to participate in writing workshop, and given regular occasions to apply their knowledge of writing to "real-life" tasks, first graders could grow and develop in dramatic fashion.

It was exciting to observe and document this change through the use of a portfolio assessment. The measure described in this guidebook allowed me to monitor student progress in a way that reflected my goals and objectives for students, and in a way that was consistent with my classroom methods and teaching philosophies. I could observe how my students were progressing as they moved from pictures to print. Were they improving in their ability to include facts and details? Could they organize their thoughts to show a logical sequence of events? Were students connected to their work through language unique to their personality? Was there an increased awareness of correct spelling and proper punctuation? The answers to these questions became readily available as I collected student work and analyzed it against some established criteria. Thoughtful analysis of this kind helped me to notice the individual writing behaviors unique to each student. I could identify specific strengths and weaknesses, and discuss them with students and parents more credibly. I could tailor each conference to the needs of the individual, and recognize common areas of difficulty where whole group mini-lessons would be beneficial. This method for rating student work allowed me to see just how powerful a quality assessment could be in informing and directing my teaching. The following serves as an example.

A View From the Portfolio: Whole Class

October

In October, I selected with each student a piece of writing to be placed in their writing portfolio. Upon examination of all the samples, I found that most students were developing nicely in their ability to generate and record a series of thoughts. There was significant improvement in spelling ability, and an expanded use of detail throughout their writing. This information confirmed for me the students were beginning to apply the methods and skills I had taught them the first six weeks of school. What I found lacking, however, was an awareness of "order," or an ability to logically sequence the events inside a piece of writing; that is, a beginning and ending with facts and details properly placed. Even in the early part of first grade we can help writers become aware of the order of things in writing. This information was useful as I planned future conferences and lessons. It suggested that students be given additional class time to review story structure and organizational patterns. Portfolio analysis provided me a deeper understanding of individual strengths and needs, and an awareness of general classroom strengths and needs; specific areas that may have, without the criteria, been easily overlooked.

I continued to see the power of this analysis and its impact on my teaching as the year progressed. Information gathered from rating student work informed me of the success of my lessons, and provided me continual direction in planning for the needs of the entire group. Monitoring class progress in all areas of writing, especially in the areas of development and correctness became easier as I looked for the application of "what I taught and modeled," to "what I saw students do independently."

January

My next formal rating of student writing took place in January. By this time, most students controlled a number of high frequency words, and were comfortable using spelling approximations to record their thoughts. At this point, I devoted several mini-lessons to story enhancement. We listened to stories where sensory detail was vivid such as *Swimmy,* by Leo Lionni and *Owl Moon,* by Jane Yolen. We described in detail events and objects familiar to our class. What did it sound like as we were walking to the gym? How could we describe the plant on the window ledge? We generated alternatives to frequently used adjectives, (nice, good, happy, etc.) and kept lists of interesting words as we heard and used them. We discussed character development, and worked to develop our skills by creating class stories that called us to use our growing knowledge. We engaged in any activity that might expand our vocabulary and stimulate our thinking.

March

By March, the time of the third rating, the portfolio selections looked dramatically different than the October and January predecessors. Students could not only generate more thoughts on paper, but the overall quality of the writing improved, and the language used showed even greater sophistication. Story leads were creative and inviting. Students often opened a piece with a question or some interesting dialogue. They took greater care to elaborate using interesting facts and descriptive details. Writing was better organized to include definite beginnings and endings. A more proficient use of spelling, grammar, and punctuation was also evident. I was pleased with this growth and felt reassured that my decisions regarding the content of my lessons were appropriate.

But while students were writing with greater ease and skill, I recognized in their writing some new areas of concern; needs that would clearly dictate my "next steps."

It was apparent that students needed continued instruction in developing proficiency with editing and correctness (capitalization, punctuation, spelling, etc.). An effort to encourage students to expand their writing to include a broader range of topics would also be necessary. Topics such as pets, family, and friends were standard and becoming routine. Students failed to see ordinary events as potentially interesting stories. The March rating provided solid evidence that my students were developing as writers, but it also indicated that students lacked variety in their work. I would need to model more writing that used daily experiences as a theme. I would also need to assign some short "free writes" to generate thinking and spark new interest. These "free writes" might include:

"My favorite part of the day is..."

"What I like to do with mom and dad..."

"Writing a letter to the school principal..."

Students were given time to generate a piece on each of these topics, then encouraged to develop one of the pieces more fully. Books from the library that highlighted routine experiences would support my efforts. I consulted with colleagues to brainstorm ideas, and looked to the "experts" (Calkins, Hansen, Atwell, etc.) for further guidance.

May

By early May, it was time for a final portfolio rating. This rating would be most exciting of all. It would provide the portfolio with an "overall" profile of student writing development; year long growth from September to May. It would illustrate the vast transition that occurs when a student moves from pictures and rough approximations, to writing that is developed and adheres to most conventional standards. It would demonstrate the power of writing as a means of personal expression and would illustrate how students develop from writing that is lifeless and artificial, to writing that is alive and purposeful. It would showcase strengths and accomplishments, and provide useful information for instruction the following year. The portfolio would become evidence of the power of writing in developing all aspects of learning.

Using Our Criteria to Develop a More Reflective View of Three Students

Andy

Andy came to first grade with limited fine motor skills. His pencil stroke was fine and wispy, and his letters were unusually large and sometimes reversed. His awareness of letters and sounds allowed him to attempt words, but difficulties with directionality, spacing and formation made reading his work a definite challenge. Andy's needs were evident and a genuine source of concern. But beneath these visible difficulties was a writer alive with purpose and originality. Andy could carefully develop his thoughts and was particularly attentive to detail. He could sequence story events in logical order and gave structure to his thoughts in uniquely personal ways. How easily I may have overlooked this hidden talent had I rated his work against traditional norms; that is, strictly precision and correctness. But analyzing his writing against a broader spectrum of criteria provided me a truer profile of Andy's overall writing abilities.

When I examined his work in the fall, I noticed how easily Andy could generate a story. His piece about sharks was several pages long and loaded with interesting facts and descriptions. His thoughts were grouped in specific order, describing first the various types of sharks he knew, and second, the characteristics and habits of each. He could sustain a topic for several days, and displayed a unique sense of style and individuality when choosing his words. Though difficulties with some writing conventions remained, improvements with spelling and motor control could also be noticed. Andy demonstrated he could write a descriptive and organized piece. What he needed was guidance and support as his muscle coordination developed. Carefully monitoring his progress through a portfolio helped me to focus on different aspects of writing. Even though the physical act of writing was awkward for him, Andy needed opportunities to write with appropriate coaching to facilitate growth and maintain self-esteem.

By the end of the school year, Andy's portfolio showcased his development as a writer. His final story about a friend was several pages long. It had exceptional use of fact and detail, and was structured in a clear and organized way. He demonstrated control over formation, directionality, and placement, and provided proof he could spell and punctuate with greater proficiency. A method for gathering and rating this work provided documentation of this achievement—where Andy began, and how he finished first grade.

Jessica

Jessica demonstrated a deeper understanding of the concepts of print. She could easily write a large core of high frequency words, and was comfortable with the process of using a pencil. Her writing was neatly arranged with appropriate spacing and punctuation. She knew that print moved from left to right, and was aware of how capital and lowercase letters were used. It was evident her knowledge of basic conventions was developed beyond that of most students her age. But generating thoughts to produce a developed and interesting piece of writing was difficult for Jessica. She was tentative in her approach, needing continual prompting to include important facts and interesting details. Her work lacked the spirit and enthusiasm unique to her creative nature. Words and phrases common to her conversational language were noticeably absent from her written work. Her pieces were short and often unorganized. She was unable to transfer her sophisticated language or her strength as a reader to her writing. A preoccupation with neatness and correctness stifled Jessica's willingness to expand her talents as a writer.

Jessica's fall portfolio selection was a story about her family. It was a perfectly spelled, properly punctuated half page of print. An opening sentence and some concluding remarks indicated a knowledge of story organization. However, the bulk of the story contained names and ages randomly placed with little thought. The story was hastily written with no attempt to elaborate on details. An internalized sense of correctness was evident, but nowhere was there the broad vocabulary and sparkling personality that invaded Jessica's conversational language.

My rating of Jessica's story confirmed my initial suspicions. She had the ability to put her thoughts to paper, but was unable to transfer her sense of story or her strong verbal skills to her writing. It seemed necessary for her to become aware of the elements of good writing and to work as a "real author" would work. Mini-lessons and conferences that encouraged the use of sensory details would be helpful. Modeling writing as a natural form of expression might also invite her to connect to her stories more easily. By winter, Jessica was writing with greater ease. Her stories extended over two complete pages and contained a broader use of description and detail. By the end of the year, she showed even greater growth in the areas of fluency, development, and organization. My early analysis had enabled me to identify specific areas of need and plan for gradual progress. Had I examined Jessica's work without a list of criteria, I may have been satisfied with her early abilities, and not planned for her development as thoughtfully.

Only through time and careful examination was I able to fully observe the true potential and contrasting needs of each of these students. Jessica, attentive to correctness and the cosmetic appeal of her paper, had the ability to easily transfer her thoughts. She lacked only the confidence and desire needed to apply her creativity and extend her thinking to writing. Andy lacked the skill and precision necessary to recording his thoughts. But his writing displayed some unusual talents and strengths not often seen in a child his age; a diamond-in-the-rough, so to speak. Having a system to monitor both student's development not only provided me specific information on how each was progressing, but also gave me a clearer understanding of my mission as their teacher. No standardized test could be as informative in noticing these behaviors, or as useful in informing my instruction.

Brian

To show more clearly how our portfolio materials can be used to shed insight into a student's writing development, let's take a closer look at one student. In Figure 3.1, Brian's portfolio samples demonstrate growth over time and provide an example of how our rating system worked to monitor his progress. His fall portfolio selection was a story about himself.

It begins with a standard opening sentence and follows with a list of the people and animals who live in his house. Brian demonstrates an understanding of some basic concepts of print (left-to-right movement, spacing between words, letter formations, etc.) He is beginning to control some high frequency words, (I, in, it, mom, dad) and is quick to notice those

Figure 3.1 Brian's Fall Portfolio Selection.

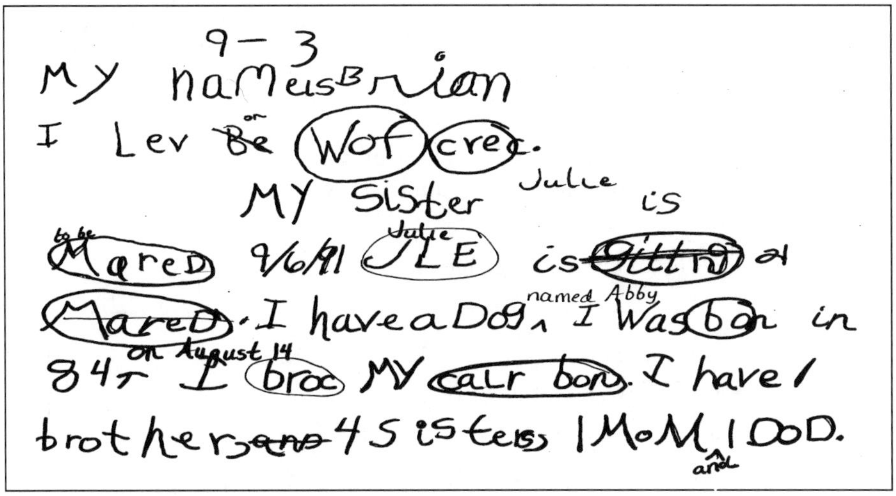

words frequently used in mini-lessons, or shared writing experiences (have, name, etc.) His sound/symbol awareness is progressing, as evidenced by his attempts to use temporary approximated spelling (crec/creek, mared/married, bros/broke, cair/collar, bon/bone). Words such as "sister" and "brother" are spelled correctly, but most likely copied from a book or other source. (Students are encouraged at this point to problem solve tough spellings. Some words become readily available as a result of this "table talk." Because stories about "family" are popular in the fall, Brian had access to family related terms.) A final editing conference with Brian helped to insert correct punctuations and sentence clarifications, a skill not yet fully developed. His fall rating was as follows:

Fluency:	4 (up to a page of thoughts)
Development:	4 (more than three pieces of information related to a topic)
Organization:	3 (limited sense of story, few facts)
Style & Voice:	2 (basic information, little description)
Editing:	3 (attempts to correct spelling, punctuation, capitalization)

Brian demonstrated in this piece that he is comfortable with writing. He can generate a story, provide basic information, and organize his thoughts by listing a series of facts. His work is typical for a child his age both in topic and development. Information is present, but description and personality is minimal. His awareness of spelling and punctuation is developing, but not internalized to the point of ease.

Brian's work had progressed steadily since the first week of school. I was confident that with regular opportunities to write, conference, and share, he would continue to progress in all areas of writing. I used the information gathered from the fall rating to plan my work with Brian in the coming weeks. We conferenced together on using more descriptive language, and on expanding his thoughts to provide greater detail and elaboration. Regular mini-lessons that modeled correct spelling, punctuation, and grammar would also prove useful in developing editing and correctness skill. And as Brian's reading progressed, so did his knowledge of story. Multiple opportunities to interact with print provided Brian the confidence to explore new topics and a growing ability to more easily record his thoughts.

By spring, Brian's growth as a writer was significant, as shown in Figure 3.2 by the writing about his cousin Steve.

Figure 3.2 Brian's Spring Portfolio Selection.

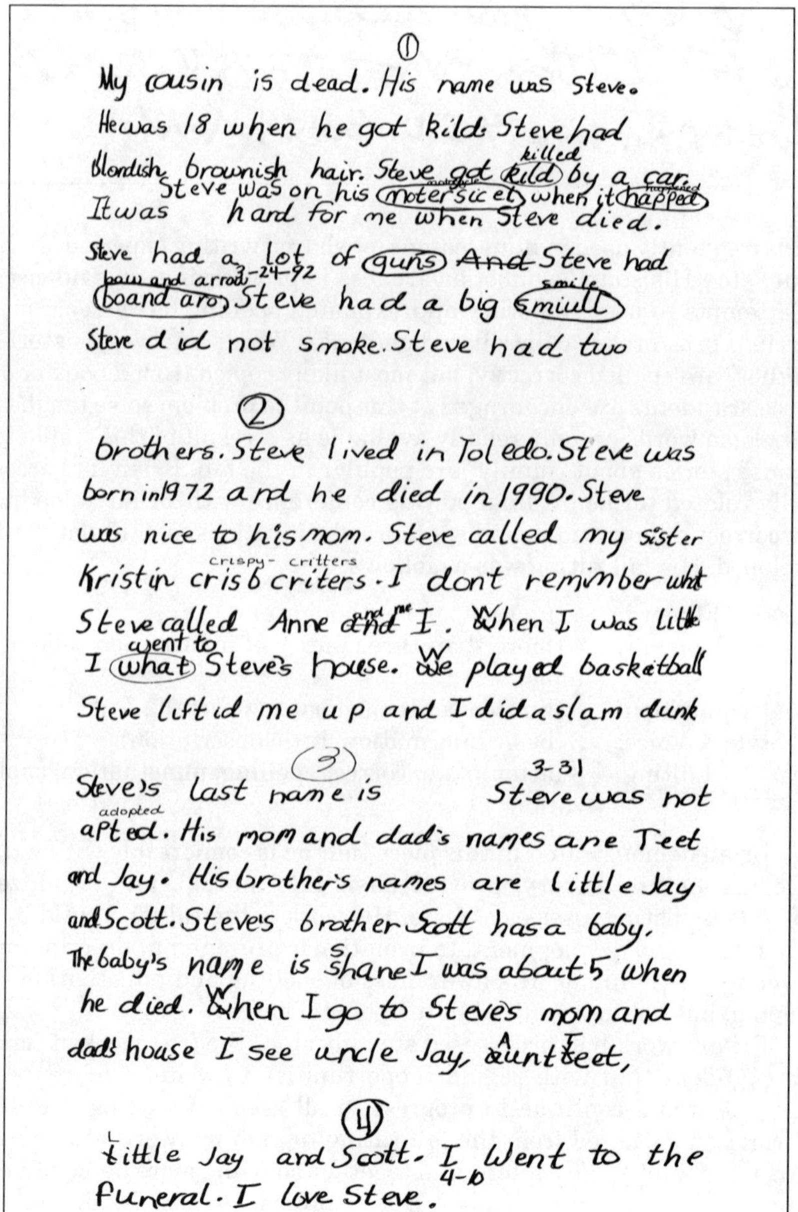

(1)

My cousin is dead. His name was Steve. Hewas 18 when he got kild. Steve had blondish brownish hair. Steve got kild (killed) by a car. Steve was on his moter sicel (motorcycle) when it happed (happened). It was hard for me when Steve died. Steve had a lot of guns ~~And~~ Steve had boand aro (bow and arrow). 3-24-92 Steve had a big smiull (smile). Steve did not smoke. Steve had two

(2)

brothers. Steve lived in Toledo. Steve was born in 1972 and he died in 1990. Steve was nice to his mom. Steve called my sister Kristin crisb criters (crispy-critters). I dont reminber wht Steve called Anne and I (and me). When I was litle I what (went to) Steve's house. We played basketball Steve liftid me up and I did a slam dunk

(3)

3-31

Steve's last name is Steve was not apted (adopted). His mom and dad's names are Teet and Jay. His brother's names are little Jay and Scott. Steve's brother Scott has a baby. The baby's name is Shane. I was about 5 when he died. When I go to Steves mom and dad's house I see uncle Jay, Aunt Teet,

(4)

Little Jay and Scott. I Went to the funeral. I love Steve. 4-10

He improved dramatically in his ability to sustain a topic, and in his ability to support his information with interesting facts and details. His voice was clear and unique as noted in his opening and closing statements about his cousin:

> "My cousin is dead. His name was Steve." (opening)
> "I went to the funeral. I love Steve." (closing)

He invites us to be familiar with his topic by including some powerful and moving details:

> "Steve got killed by a car. He was on a motorcycle when it happened...It was hard for me when Steve died." "When I was little I went to Steve's house. We played basketball. Steve lifted me up and I did a slam dunk."

Brian is involved with his topic as a writer. There is feeling and emotion throughout the story. He has learned the power of the written word as a form of communication and expression. An assessment of this piece illustrates this improvement:

Fluency:	5 (three or more pages)
Development:	5 (exceptional use of fact and detail)
Organization:	4 (opening sentence, details, conclusion)
Style & Voice:	3 (attempts sentences consistently and uses some descriptive language that shows personal involvement)
Editing:	5 (evidences internalized sense of correct spelling, punctuation, and capitalization at age appropriate level)

Final Thoughts

The way this form of assessment works to monitor progress and diagnose "next steps" is of far greater benefit to me than the standardized achievement tests or "single piece" writing evaluations that continue in our schools. While these traditional measures seem here to stay, and while they may provide some information regarding a student's ability to answer correctly on a given day, the amount of emphasis given to the content of these tests, and to preparing students to take these exams is troubling at best. Often, authentic learning experiences that prepare students to independently acquire, use, and monitor their learning are passed over in favor of experiences that involve students in tasks that are patterned after the state's yearly assessment measures. Even those who are sensitive to what research calls "best practice," fall victim to the pressures of "doing well on the test." And so we lose sight of what teaching and learning should be about, that is, involving students in the kinds of meaningful experiences that allow them to not only attain skill, but to use their skill to create deeper and richer levels of understanding.

It may be understated to say that I am excited about portfolio assessment and its influence on teaching and learning. Nothing has had a greater impact on my professional life than my involvement with portfolio analysis. It is exciting to see my students prosper as writers and to know that my pursuing the portfolio discipline helped foster this growth. Through it, I could recognize student strength and deficiencies and plan for individual needs accordingly. The procedure eliminated the "guess-work" from my planning, and allowed me to effectively focus my resources. What at first seemed like "hard-work," in reality became "smart-work."

I continue to use the materials outlined in this guidebook and support the efforts of others who value this work as well. I find it to be a reliable assessment tool, and a necessary resource in guiding young writers in the classroom toward greater levels of proficiency.

Chapter 4

Improving Our Teaching Through an End-of-Semester Review: Upper Elementary

Aaron Schippert and *Alan Willoughby*

Why Rating Scales are Valuable for Teacher Growth

When we began teaching writing with a workshop approach several years ago, one of our main goals was to get children to write. As we provided time, choice, a familiar structure, opportunities to respond to writing, and a supportive community, students began to see themselves as writers. As pleased as we were to have students enjoying writing, we were only at the beginning of a long journey. Over the years we have developed our practices to help our young writers improve the quality of writing by focusing on issues such as beginnings, endings, character development, writing in different genres, and correctness issues such as editing and spelling. Students regularly evaluate their own writing for quality. Of course, reading and responding to literature have been instrumental components in guiding us to higher standards of writing. Students continue to improve in the difficult task of learning to be effective writers.

How do we know that students are improving as writers? How do we know that our teaching practices are effective in guiding young writers to improve their skills? Teachers can easily discuss a student's collection of writing at different points throughout a year, although that isn't often done, and generally determine that a student is growing as a writer. But what exactly are we noticing? We don't often hear specific conversations about the nature of the improvements, or spend the time to carefully analyze the work with a set criteria. Are we really capable of doing our best as writing teachers without taking a sustained, focused, collaborative look at student work?

It is these questions that guide us in our next stage of development, and the criteria for evaluating student writing created through the Lenawee County Portfolio Project are our reference point as we enter this stage.

In a chapter titled "Making Assessment Work for Students and Schools," from the book *Authentic Assessment in Action: Studies of Schools and Students at Work*, Linda Darling-Hammond and her colleagues state:

> By working on standards and assessments, looking carefully at students' work and progress, and working to develop supports for student success, teachers are engaged in constant learning about the needs and talents of their students, about the effectiveness of their teaching strategies, and about the nature of teaching and learning. By working toward challenging standards on authentic tasks, students are engaged in constant learning about the nature of high-quality work, about themselves as learners and workers, and about the phenomena they are studying (p. 252).

We think this defines the direction in which we need to go to reach another level of quality in student work and our work as writing teachers. Further, as we develop in this new direction, it will be important to spread the opportunity to other teachers in our building and throughout our district.

As the authors of *Authentic Assessment* also point out:

> The process of developing assessments and evaluating student work is, in itself, a powerful form of professional development—one which recognizes that growth results from teachers engaged together in dialogue and reflection rather than from 'experts' who deliver information at training sessions or one-shot workshops (p. 254).

How We Began Looking at Student Work

For several years in our district, Hillsdale Community Schools, we have collected students' writing work in a portfolio. We collect the work for purposes of student self-evaluation, to look at writing growth over time, and to show parents. These are worthwhile purposes of a portfolio collection. However, we have often felt that to achieve all of the above as fully as possible, and to more fully inform our teaching practices, we needed to have a set of criteria that would allow us to focus on issues of quality writing. The idea of having a clear and specific set of standards for student writing seemed to have the power to create a common language of quality writing that could be shared by teachers, parents, and students.

A few years ago our district held a class to study the use of portfolios, directed by Dick Koch of Adrian College, the Director of the Lenawee Project. At that time we were introduced to the rating criteria, and we talked about providing ourselves the opportunity to use the criteria to study our student writing portfolios. The class was well attended, and there seemed to be genuine enthusiasm for the work, and yet we failed to follow through with what was started. Perhaps the group was too large and we did not have a common vision of what our next step in improving our writing program should be, or of the value of portfolios. So we all went our separate ways and continued to feel that we knew children were making progress, even though we were not specific about the nature of the improvements.

Fortunately, we have worked fairly regularly with Dick for several years. The idea of the rating criteria was reintroduced through several conversations, and this time some of us were receptive to the possibilities that focused collaboration could bring. It is hard to find time to get together in the school schedule with the increasing demands of curriculum. However, we knew this was important and long overdue. We set up a regular once a week schedule to meet after school and go through student writing together. We gathered student portfolios, the rating criteria, and began to analyze the students' writing and our teaching practices. Using the criteria and rating scale values we hoped would give us a clearer picture of the kind of growth students were making, and of the effectiveness of our writing programs.

We began looking at a fifth grade student's work, and it was interesting because we both knew the student from our classrooms. It was also satisfying to see the progress he had made from third grade. We read four of John's writing pieces and discussed each in terms of the criteria for fluency, development, organization, style and voice, revision, and editing and correctness. With the common language and focus that the criteria provided us we were able to see John's work in new ways. Discussing it together was valuable in clarifying the criteria, and in bringing to light the context in which the writing was done. Sometimes the fluency category was questioned because students were engaged in a certain type of writing or because a particular time limit had been set on an assignment. Those conversations and compromises were valuable in understanding John as a writer, and ourselves as teachers. As we evaluated the work, several patterns were revealed which would not have been available to us had we stayed with our previous practice of generally seeing improvements. All of John's work is about the same length, approximately one page, so his fluency rating remains at a "1" or "2." He is beginning to develop his writing with details, but this too is an area he can improve on. His organization has shown steady improvement (1 to 3). John's voice is becoming more evident (1 to 3). Revision needs to become a focus for John, perhaps this would help fluency and development. He is very strong in his editing skills.

We could see that John needed to have some guidance and coaching to improve his fluency, development, and revision. Knowing this alerted us to plan lessons and to provide examples from other authors for students needing improvement in these areas.

We felt that our time was well spent after one meeting. We had learned a great deal about one writer, and we were formulating questions about how to help John improve as a writer. We discussed the issue of fluency, whether it is teachable, and how to help students to become more fluent. We realized that we had not been thinking directly of fluency as we presented our lessons, and yet there were things we were teaching that would help a writer develop fluency. What were those things? Why were some students, like John, still having trouble with fluency? These were only some of the valuable questions that came from our first meeting. We continued to meet and look at several students' writing. We became pretty comfortable with the rating scales and felt that our focus was getting clearer. We began thinking about the different lessons that we could do, and that we already do, to help students achieve greater quality in their writing. We didn't think it was necessary to rate all the writers in our classes together. It is hard to find the time, and now we were confident that we could analyze our own students' writing. However, it is still valuable to get together to rate a few portfolios in order to share information about the context of pieces, discuss possible strategies, and share successes. Each time we get together to discuss student work we become more aware of our students as writers, and what we are doing to help their development as writers. We are learning what each other focuses on most during the writing period and what resources we use in our lessons. It has been an excellent professional development experience to meet together and examine both student work and our own practices.

Alan's Work with Third and Fourth Graders

I have a third and fourth grade split classroom this year, and one strategy that I have in place to be able to manage this situation is the workshop approach to writing. It allows students to develop at their own rate and practice their writing skills in a supportive community. It allows me to see that some third grade writers have more highly developed skills than some fourth graders, and that writers of all ages have their strengths and weaknesses. Using the rating scale helps me clarify the nature of the strengths and weaknesses, and allows me to develop lessons to help all the students in my class.

Fluency

Casey is a third grade writer who enjoys writing. He has written two nonfiction pieces, one on the origins of Halloween and the other about a friend's party. In addition, he has written an essay on the importance of being a good pet owner and a poem. Casey is a fluent writer and I scored him a "4" on "Halloween" because he wrote more than four pages quite easily. However, since this was a report style of writing, he gathered information from books and tried to put it into his own words, something that is difficult for third graders to do. So, he may not have achieved this same rating for fluency on another piece of writing. On his piece "Zac's Party," I rated him a "5" for fluency because he wrote more than four pages with flow throughout, and represented his own thinking. His essay on pet ownership would receive a "2" on the rating scale because it was a little over one page in length. However, that was the expectation for this assignment. I asked students to write a tightly controlled piece to develop their thinking and organizational skills.

Many students had a difficult time writing smoothly within the confines of an essay format. Is Casey a fluent writer? I would say that he is fluent with some types of writing, and he is less fluent with other types of writing. He is comfortable expressing himself and communicating effectively with some kinds of writing. I think most of us adults would agree that we are much more comfortable with some kinds of writing. The scale served as a useful framework from which I could examine and think carefully about what fluency means for different genres explored in my classroom. Fluency has something to do with the number of pages written, depending on the genre. However, I have come to realize that it primarily involves the ease and flow with which the work is written. The rating scales have helped me think about fluency in this new way. Doing the analysis also leads me to question why most children in third grade are most fluent writing personal narrative and fiction. Of course the answer is that those kinds of writing are what they have practiced most. This is what many students have mostly read, too.

When I try to expand students' ability to write effectively in essays, letters, and other forms of nonfiction, such as history, I should expect less fluency. One method that I use to help students develop fluency in any genre is to model my own writing, and examine with them the particular type of writing by reading lots of examples written by professional authors. In order to help students with essays, I wrote my own essay on car safety. We also read essays from our local newspaper and another newspaper. After this modeling and examination of essays, along with our discussion of the purpose for writing essays, students were better prepared to choose topics that were meaningful to them and better prepared to write more fluently. Also, having students share their essays, drafts and completed pieces, and then discussing them with the class, gives us more experience with this kind of writing. An excellent book that I use to help myself understand and write essays is *Crafting* a *Life in Essay, Story, Poem* by Donald Murray. I even read some of his advice to the class, although the book is intended for a much older audience. I think it is extremely important to share my learning with students, and to invite them on an exploration of writing with me.

Another way to help children achieve greater fluency is to develop a checklist of criteria for the particular type of writing you are developing such as the example checklist in Figure 4.1.

For example, I expected a good essay to include an engaging beginning, a paragraph to put the topic into context for the audience, support or examples for the author's ideas, and an ending that helps the reader to think in a new way about the topic. Of course, I had taught lessons to help students understand each aspect of the essay as I mentioned earlier. As the class is working through the different criteria that should be present in a piece of writing, they evaluate their work and better understand how to write in a particular genre. It is very helpful to allow students a second or third chance to write in a genre in order to try independently writing that genre with the criteria in mind. I see students struggle with essays the first time, and then as they gain experience with the framework of the checklist they become much more comfortable, or fluent, with them. It is particularly difficult for young writers to find a way to put their topic into a broader context for their readers. Also, it is hard to move people to think in a new way with the ending. Several students struggled with these aspects of the essay, so I had individual conferences to help them find a way

Figure 4.1 Fluency Checklist

Third Grade Journey Sheet

Author: ______________________________

Title: ______________________________

Genre: Essay ____________________

	Me	Friend	Mr. W
1. Is the beginning personal?			
2. Did you put the topic into context?			
and invite the reader?			
3. Did you give examples or reasons to support your topic?			
4. Does the ending make the reader think?			
5. Is your essay organized effectively?			
6. Did you use paragraphs?			
7. Are "No Excuse" words spelled correctly?			
8. Did you use correct punctuation (periods, question marks, capital letters, etc.)?			
9. How did you choose your topic?			
10. What did you do that is new?			

to effectively do these things with their own particular topic. After these conferences, and many student essays shared with the class, students began to think more carefully about the criteria of a good essay as they wrote their drafts. It may not mean that they are writing more pages, because that is not the emphasis of the essay criteria, but what they write is done with more ease and flow. It makes sense that the more students practice, the more improvement they will make.

Development

After rating several students' writing it is possible to see patterns in the categories of the rating scale. I can see that many students develop their pieces with facts and some sensory detail (a 3 or 4 on the scale) and writing is fairly well organized with logical sequence and arrangement (a 3 to 4 on the rating scale). However, I notice that very few students have developed in the category of style and voice past a "2" or "3" rating. Also, revision is an area where they need to become more effective. This kind of analysis is extremely valuable in informing me of my strengths and weaknesses as a writing teacher. What is it that I have done to help students organize and develop their writing? What is happening at other grade levels to promote organization and development? Why are students not as strong displaying their style and voice, or revising their work effectively?

Many young writers are not aware of their audience. Therefore, they tend to leave out many important details that would be helpful to the audience. As I look at Nick's work from the end of last year to this year, I see improvement in the development of his writing. He is more aware of his audience and writes with more details and elaboration of facts as a result. At the end of last year Nick wrote a piece called "Mr. Shoe" (Figure 4.2) in which there is some sensory detail, but much of the action is quickly stated and not well developed.

Figure 4.2 "Mr. Shoe"

Mr. Shoe

Once upon a time on April 13 there was a man named Mr. Shoe. They called him Mr. Shoe because he had lots of shoes.

In the morning he was playing with a shoe and slipped on a banana. It hit his tv and flew in his mouth fast as a bullet.

He wanted to take a walk but he opened the closet and "AHHH!"

Boom! All the shoes fell down like a pyramid not put together. "Ow!", said Mr. Shoe. It took 1 hour to pick up the shoes.

In his piece this year about NASCAR racing (Figure 4.3), he has his audience better in mind as he provides more elaboration of facts and details.

Figure 4.3 "NASCAR"

"NASCAR"

NASCAR is a sport that people race at racetracks. Their cars are designed to race at a track. As the crews are at the track they work on back-up cars for the next race in case there is a crash. It can be very dangerous for spectators and the drivers.

The crews need lots of practice for the pit stops. For the pit stop they need six guys over the wall to do different jobs. There is a front tire changer, a back tire changer, a jack man, a gas man, and two tire holders.

It can be dangerous, racing at 170–200 miles per hour passing other people. Some people flip their cars, and they can spin their car into the wall. Some people want to win so bad they get going so fast they blow their engine. If someone rear ends another person they can make their gas tank explode! The safety crew helps the driver get out of his car or puts out any fires that are in the car. The drivers stop, drop, and roll if they're on fire.

I had noticed this improvement in development in general, but with the rating scales I have been able to think about the development in more specific language and concepts. Doing this helps me analyze what lessons have led to this growth in writers. There are several things that I do to help students develop their writing with sensory details and elaboration of facts. The first writing I have my students do at the beginning of the year is in a writing notebook. The notebook is a place to keep observations about their lives. We work on entries that include memories, observations, wonderings, stories that are associated with important objects, family stories, and others. I help students write with as much detail as possible in order to become keen observers of life. We share our entries often to get new ideas and listen for examples of good writing. In order to introduce a particular kind of entry, I read aloud a picture book that relates to that type of entry.

I read *Letting Swift River Go,* by Jane Yolen, to introduce entries about memories, and *The Table Where Rich People Sit,* by Byrd Baylor, to talk about objects that have special meaning and the stories behind them. *I'm in Charge of Celebrations* and *Guess Who My Favorite Person Is?,* both by Byrd Baylor, are excellent books for helping students learn about using details to describe things. There are many wonderful picture books with which to illustrate development of writing. Some excellent references on writing notebooks and helpful picture books are *Lasting Impressions,* by Shelley Harwayne, and *A Writer's Notebook* and *What a Writer Needs,* both by Ralph Fletcher. They have helped me understand the value of notebooks and helped me develop my own notebook. I always share examples of my own when introducing a kind of notebook entry. After taking these authors' suggestions for books to share in order to help develop the use of notebooks, I now have added many more books to my list of read alouds for the purpose of demonstrating a notebook entry. Students practice developing their writing with detail regularly through the use of the writing notebook.

Another way I help students develop their writing is by looking at good examples of development from chapter books that I have read to students. I might read a paragraph from a book by Natalie Babbitt, such as *The Search for Delicious,* without the detail that she uses and then put the page she wrote on the overhead projector in order to compare the difference. After discussing the detail in the paragraph I ask students to each find an entry in their notebook that needs development. They add details to develop the piece so as to more fully involve the audience in their writing. We share the before and after versions and discuss how the development worked, or could still be improved.

Organization

As I examine the organization of my students' writing I observe that at the start of the year most students are capable of a "2" on the rating scale. They have a general sense of the proper sequence of a piece of writing. From there many students consciously start thinking about putting in an effective beginning and ending. In order to help students begin to reach for those goals I present lessons on how to begin and end. Of course, beginnings and endings, too, will be different for different kinds of writing, but all writing requires an effective beginning and ending. Fletcher's *What a Writer Needs* has clear information on these components of good writing. His helpful suggestions and extensive list of books that serve as examples of the different aspects of effective writing have been invaluable to me as a writing teacher.

Chris Van Allsburg has many books in which he uses a surprise ending. I read several of his books and discuss endings with the class. I read aloud several other books that have different kinds of endings too. Students analyze books that they have read and discover what kind of ending their authors have used. I then ask students to try several different endings for the writing they are working on, and we share these with the class. As students do this they get involved in debates about the merits of different endings for their writing and increase their understanding of organization. The same kind of lesson can be done with beginnings. Casey came up with four different beginnings for his piece on Halloween. He then chose the one that he thought was most effective at inviting his audience to discover something about the origins Halloween.

The work with the rating scales has helped me evaluate the usefulness of lessons that I believed were successful, but now I have specific indicators of quality writing to support my beliefs.

Aaron's Work with Fifth Graders

An experience I had with a student who might seem invisible in many classrooms can show why I value our rating work with upper elementary students. It was early fall and I really didn't know my students the way I would later in the year. There was a shy little redheaded fellow named Chad who sat quietly in the back corner of the room. When I sat down with Chad to teacher conference with him on a writing that he had considered publishing, I strained to hear as he somewhat mumbled the words of his piece aloud to me. Slowly, as Chad labored through his story on Cedar Point, a nearby amusement park, I began to understand more and more about this child. The piece showed that in writing Chad could achieve powerful use of voice. He was hilarious! He could sustain an idea for well over two pages, and although he lacked the use of certain writing conventions, he had the ability to write in full and complete sentences. The piece showed a beginning, middle and end with a style completely unique to the author. This shy, unapproachable young person was able to let his ideas blossom on paper because he is a product of a writing process school that allowed him the time, structure and opportunity to write and write well.

In the past, as much as we knew about the writing process, finding matching assessments had been difficult. What we needed to do more than anything was to investigate new assessments and figure out how we could use them in our school and district. About the time that we began investigating assessments I came across a book called *Expanding Student Assessment*, edited by Vito Perrone. In it Ed Chittendon claims that teachers and districts need to find out for themselves how assessments are going to work for them:

> In an assessment program teachers participate in a common plan of data collection and review, which might be called a plan for documentation. Guidelines are followed so that the data are sharable, public and open to examination. But the valuative judgments concerning the implication of those data—whether the judgments pertain to the progress of a child or the quality of a program—are necessarily more complex and open to debate and discussion (pp. 26–27).

I found this particular passage to be extremely useful. It was basically telling us to collect work, evaluate it, then estimate the implications for further teaching decisions.

Collecting and Evaluating Work

The first order of business would be to accumulate student work by building a writing portfolio. Getting student work would be no problem. From the very first day of writing workshop, I ask my students to save and date everything they write. The dates on the work really help the child or teacher determine when a skill or writing convention has been mastered. Also, when the students save all work, we can see what has transpired between a first and second draft of writing, or between a second draft of writing and a published piece. This collected work can then become a basis for students to self assess so they can more easily continue to grow.

I also use the portfolios for Student-Led Conferences. Student-Led Conferences are an alternative to traditional Parent-Teacher Conferences. Instead of the teacher talking to parents about their children, I have the students talk to their parents about their schoolwork. I believe the students can explain their work and thought processes better than I can. Students collect early samples of work (writings and reading responses) and compare them with samples from later in the year. The presentation they then offer to accompany the work gives them every opportunity to share with parents what they're doing well in and what they need to continue to work on. Although I liked the Student-Led Conferences and felt that they had some merit in my classroom, I didn't feel that they were a comprehensive assessment tool. What I really needed was a system where I could look at student work and have, in Chittendon's words, "the evidence associated with . . . indicators . . . be unambiguous to the extent that parameters of its collection are understood by teachers, parents, and students" (p. 27).

As Alan indicated for us, that system would come from our mentor for integrated language arts—Dick Koch of Adrian College in Michigan. The grade five rating values sheet included a "1" through "5" rating for six different writing categories including fluency, development, organization, style and voice, revision, and editing and correctness. With Alan's help, I took the fifth grade rating scales and applied it to five students in my classroom. I chose three writing pieces from each student that represented an early piece of writing, a mid-year piece of writing, and a later piece of writing. We read the pieces together and tried to make sense of the rating scales as they applied to the writing. It didn't take long for us to agree on our scores for the papers.

Style and Voice

One early pattern I found was that the fifth grade writers had the ability to write with what I considered good style and voice. That is, the students seemed to demonstrate mostly effective word choice and showed the ability to write in a way that conveyed who they were as individuals. Below are samples of two writers Alan and I rated a "4" in the style and voice category.

Kayla is one of my fifth graders who I also happened to teach last year in my fourth/fifth split class. I would characterize Kayla as a risk taker with her writing. She was always willing to try to apply new mini lessons in her writing. For example, if I had recently taught quotation marks for dialogue, Kayla would immediately try to use them in her piece. I think Kayla has the ability to use writing conventions in a way that is unique to her style and shows her personality as she writes. In her piece "My New House" she demonstrates this.

My New House

"Mom, tomorrow we don't have school, right?" asked Kayla.

"Yeah, so," said mom like she didn't really care.

"Well, can we go to the mall and go shopping?"

"Well, since the semi is not here I guess so, but before you get all happy, go ask your dad."

I ran from the kitchen where my mom and I were, all the way through the hallway and into the bathroom where my dad was combing his hair.

> "Dad, canwegotothemall," said Kayla out of breath from running and talking so much.
>
> "Za zoo za zoo. That's what you sound like. Talk slowly," said my dad like a monkey.
>
> "I said can. . . . we . . . go . . . to . . . the . . . mall . . . .and . . . go . . . shopping?" asked Kayla mocking her dad.
>
> "I don't know, go ask your mom," said dad thoughtfully.
>
> "Forget it. I'm not running back and forth three million times," said Kayla.

Another example of strong voice comes through in Nadia's piece called "The Car Ride to King's Island." I had done some lessons on characterization, and I assigned a writing task in which I asked my students to tell what their characters say and do as a way for the reader to get to know them better.

> The Car Ride to King's Island
>
> "What's your name?" I asked Shane's friend.
>
> "Cole." He said it like he was in a very important conversation with Shane and did not want to be bothered.
>
> Yeah right, the only thing two teenage boys could talk about would be girls.
>
> I'll tell you a secret—that's the only thing they did the whole trip was look for girls, but they didn't find any.

Editing and Correctness

Another rating category where the students scored very well was editing and correctness. It's probably important here to note that correctness is anything but an issue during the initial drafting phase of writing. I strongly advise my students to first get on paper all of their thoughts on a particular issue. Often, that means there will be plenty of punctuation and spelling mistakes on early drafts. That is perfectly fine because I know the students will be sharing all of their thoughts and not getting bogged down when trying large vocabulary words and not spelling them correctly. Similarly, getting the exact punctuation mark at the moment of writing the first draft is not needed because my students are allowed to return to a piece several times to make the necessary corrections when they choose to publish it. Timely mini-lessons can add to student learning about correctness and editing issues.

Spelling immediately comes to mind when we think of correctness. Without fail we could look at young writer rough drafts and find plenty of spelling blunders. This is not, however, a time to despair over the "troubled spelling of American youth." Rather, it is a golden opportunity to have a meaningful discussion on how young spellers can make corrections on their pieces. One thing I like to do is have the writer circle any word that he *thinks* might have been misspelled as in Figure 4.4.

Figure 4.4 Self-corrected Spelling Sample.

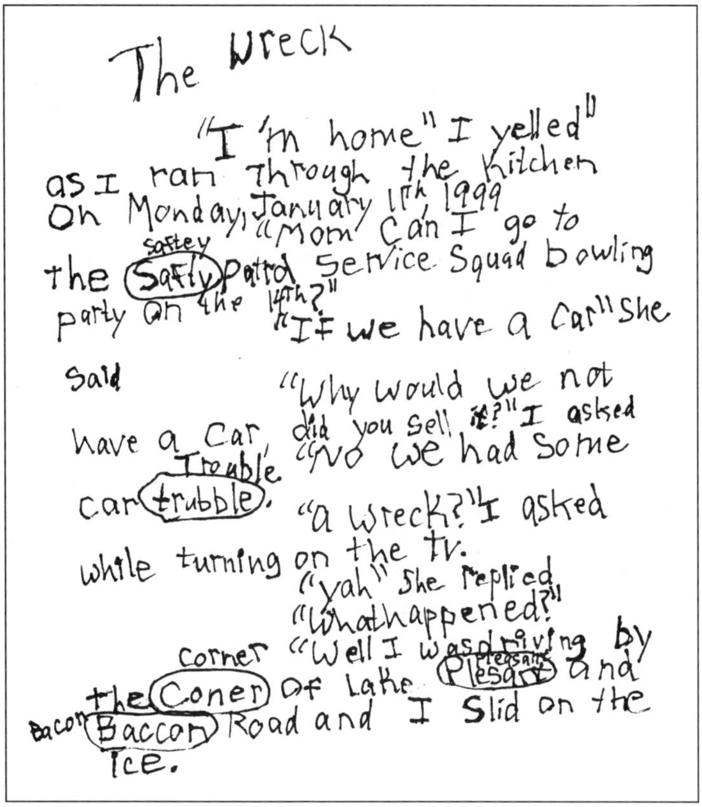
The Wreck
"I'm home" I yelled"
as I ran through the kitchen
On Monday, January 11th 1999
"Mom can I go to
safety
the (Safly) Patrol Service Squad bowling
party on the 14th?"
"If we have a car" she
said
"Why would we not
have a car, did you sell it?" I asked
"No we had some
Trouble
car (trubble). "a wreck?" I asked
while turning on the tv.
"yah" she replied
"What happened?"
corner "Well I was driving by
Pleasant
the (Coner) of Lake (Plesant) and
Bacon (Baccon) Road and I slid on the
Ice.

It's amazing to me how many times the students will circle words that are in fact misspelled. This tells me that they are not far off from spelling the words correctly. From here, we can take our list of circled words and simply write them two or three different ways. Often, the correct spelling will be in the last attempts, and the writer can identify this either alone or with the help of a friend. And certainly, if the word still looks wrong, they at least have a better choice as they refer to a dictionary. As the student gets closer to the publication stage we shoot for no spelling mistakes.

When I teach a series of lessons on paragraphing, I like to use the books that my students are reading in reading workshop. I'll ask the students to write down examples from the book they are reading that illustrate good use of paragraphing. This will lead to a great discussion on what published authors do when moving a story along. Often we'll see how an author found a natural transition from one idea to the next and made the shift into a new paragraph. The students also notice that when a different person talks in a story that requires a new paragraph.

We'll take all this information and apply it to a child's piece of writing. I'll get permission to take a student's rough draft, make a transparency and put it on the overhead screen. This way the whole class can see how we edit a piece for correctness. Working on paragraphing as a class we'll determine where the natural breaks in the piece occur. I teach the students the paragraph symbol, and we'll include that where it belongs throughout the piece. Next, we'll rewrite the draft with proper paragraphing so the whole class can see what we've accomplished. After doing this a few times, the students get a good feel about how to paragraph, and they become better and better at it.

Revision

As Alan and I looked closely at student work it became apparent to me that my students were coming up short in the revision process. Revision, as Jan Turbill describes in *Now We Want to Write!*, "is when the writer adds, deletes, changes, or rearranges words and parts of the writing . . ." (pp. 34–35).

It might be important here to note that I accept that learners come to me with varying degrees of ability and mastery. An upper elementary student might be expected to know a range of skills and tools as a writer, but certainly some students will know more, others less. I believe that, as Cambourne and Turbill suggest, I have to "accept approximations" in my students' efforts. However, using the rating values showed me that, even though I had taught some lessons on revision, I needed to teach more.

Sometimes, selling fifth graders on the notion of revision is not easy. Some young writers (and older ones too) like to believe that a first draft *is* a publishable draft. One way that I combat this tendency is by using award-winning author Gary Paulsen. I read aloud to my students every day for many different reasons. Early in the year I'll intentionally read a Paulsen title like *Bryan's Winter* or *Hatchet.* Paulsen's books are extremely popular in fifth grade, especially for the boys, and generally many students will get turned on to reading because of this. Without fail, about five or six students will get another Paulsen selection when I finish my read aloud book. Here now is my opportunity to teach revision techniques in writing.

A couple years ago Trumpet books produced a series of videos that highlight children's authors. In one video Paulsen talks about how he loved to write but his publishers were always telling him he had to rework and rewrite different parts of his stories, and he hated that. That is, he hated that part until he found computers and found that you could cut and paste and do all sorts of neat things with your writing. I use that part of the video to help me explain that all writers have to revise, and that our best pieces come as a result of revision. I think my students can identify with an award-winning author, and, therefore, they feel they are not alone when it comes to revision—we all have to do it.

To increase my work with revision, in light of Alan's and my reflections, I did what I always do when I need ideas for teaching mini-lessons—I referred to the experts. In this case, I referenced the excellent book I mentioned above by Jan Turbill. In a chapter called "The Writing Time," she explains some teaching strategies for different craft issues that have worked for her over the years. One of the strategies she presents that I always teach early in the year is on how to achieve effective introductions (p. 37). We look for alternatives to the "Once upon a time . . .", or "One day

I was . . .", type lead. One great place to go for alternative introductions is our class library. We have a wide range of award winning books and authors to choose from, and I encourage the students to reference these materials for ideas on introductions. I'll read the first few pages of Avi's *Wolf Rider* as a lead in for a mini-lesson on "shock intro's." It reads:

> The kitchen phone rang three times before Andy picked it up. "Hello," he said.
> A voice replied, "I just killed someone."
> "What?"
> "I just killed someone."
> "I don't understand," said Andy.
> "I have to tell . . ."

That introduction gives the class all sorts of ideas about what might happen in this book, and we discuss why it is a clever introduction to a story. It invites people to read because they are shocked and made curious by the opening.

Another favorite of mine is *Shadow of a Bull* by Maia Wojciechowska. The story begins like this:

> When Manolo was nine he became aware of three important facts in his life. First: the older he became, the more he looked like his father. Second: he, Manolo Olivar, was a coward. Third: everyone in the town of Archangel expected him to grow up to be a famous bullfighter, like his father.

Again this is a powerful introduction and I use it to inform the class that the best kind of introductions are those that capture our interest immediately. This is a very clever introduction in that the boy is expected to be a great hero like his father and yet he is a coward. The author has managed to capture our interest in just a few short lines. It makes the reader want to continue reading the story.

Finally, Karen Cushman intrigues us with her opening to *The Midwife's Apprentice*:

> When animal droppings and garbage and spoiled straw are piled up in a great heap, the rotting and moiling give forth heat. Usually no one gets close enough to notice because of the stench. But the girl noticed and, on that frosty night, burrowed deep into the warm, rotting muck, heedless of the smell.

Karen Cushman arouses our sense of intrigue. What kind of a character could she possibly develop who can endure an introduction such as that? This becomes the topic of great discussion in my classroom, and, more importantly, the students are thinking of alternative ways to introduce their stories. After sharing excellent introductions like the ones above, it doesn't take much to get my students trying new and interesting ways of beginning their stories.

There is a revision strategy built directly into the writing process itself. It is the friend conference portion. During this time a student who has a piece of writing they consider publishable will read it aloud to a friend to get feedback on the piece. They have been coached in advance, and follow the steps listed below during the conference.

1. Tell the writer what you really like or thought was interesting.
2. Tell the writer what you thought he/she did as a writer that was good (style).
3. Tell the writer what you would like to know more about (omissions).
4. Make a practical suggestion on how the writer might improve the piece.

The writer is encouraged to write these ideas down on his journey sheet and to try to find a place to use them in his piece as he rewrites.

Sometimes the best lessons in reading and writing strategies come directly from the students. I was recently reading one of my student's stories when it occurred to me that the piece would make an excellent lesson about a gap in writing. By "gap" I mean the child left out pertinent information in the story that would hinder the readers' understanding. There were also some key elements of the story that were out of sequence. I just knew this would be a perfect opportunity to use actual student work for a mini-lesson on revision dealing with omissions and sequencing. I had gotten permission from the writer to use her piece for the mini-lesson the following day, and I planned that as a class we would help her to see the omissions in her piece (Figure 4.5). Before I even got started, the little girl, Kayla, had told me that she had already revised her piece based on the suggestions of another girl, Lauren, with whom she had a friend conference. I had picked up the first draft of "My Sister's Accident" when Kayla shouted, "Mr. Schippert, you don't have to share that with the class. Lauren helped me."

"What do you mean?" I asked.

Lauren interjected, "Well, she means that yesterday we had a friend conference and I helped her."

As it happened, I had overheard their previous conference. As soon as Kayla was done reading Lauren had some questions. "Who do you mean when you say 'they left'?", Lauren inquired.

"Oh, my sister went to Jackson with her friends, Valerie and Erin."

So, the first bit of advice that Kayla received from Lauren was to tell the reader who is in the car and who is driving the car. Later, Lauren asked Kayla, "Why do you say 'my mom left right after they did'?"

Immediately, Kayla realized that part of her piece wasn't necessary. Lauren also noticed that there was a lot of superfluous information about Kasey in the first paragraph, and she pointed that out to Kayla, as well. The last point Lauren made was about driving on "12." She suggested that Kayla explain to the reader that "12" is a road. This actually made Kayla realize that U. S. 12 was not the road that the accident happened on, but it happened on Milnes Road instead. After Lauren's and Kayla's friend conference, Kayla cut out unnecessary information in her earlier draft and added more useful facts and details as shown in Figure 4.6.

What happened between the girls was just too good to let go. I immediately had the girls share in front of the class what they had accomplished

Figure 4.5 Gap in Writing Sample.

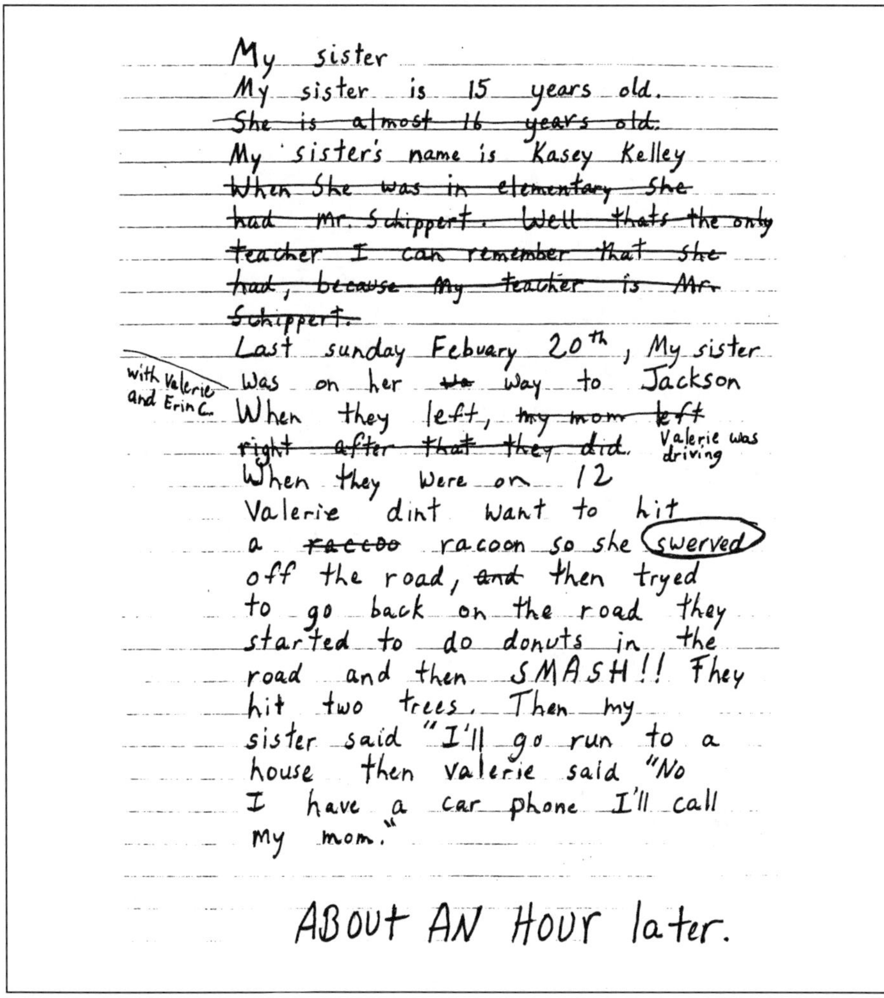
My sister
My sister is 15 years old.
~~She is almost 16 years old.~~
My sister's name is Kasey Kelley
~~When She was in elementary She had Mr. Schippert. Well thats the only teacher I can remember that she had, because My teacher is Mr. Schippert.~~
Last sunday Febuary 20th, My sister Was on her ~~w~~ way to Jackson with Valerie and Erin C.
When they left, ~~my mom left right after that they did.~~ Valerie was driving
When they were on 12
Valerie dint want to hit a ~~raccoo~~ racoon so she swerved off the road, ~~and~~ then tryed to go back on the road they started to do donuts in the road and then SMASH!! They hit two trees. Then my sister said "I'll go run to a house then valerie said "No I have a car phone I'll call my mom."

ABOUT AN HOUR later.

during conference time. The class got a better lesson than I ever could have taught. They saw firsthand how they might become models for a particular strategy.

Connections to State Assessment

The state of Michigan requires all of its fifth grade students to take a series of MEAP tests (Michigan Educational Assessment Program) every February. Included in these tests is the writing proficiency assessment, where the students have three, forty-five minute sessions to complete a published draft of writing on a given topic. On the back of the rough draft booklet that the state provides for each student is a holistic scoring rubric. The state gives a score based on how well a child can produce a piece of writing meeting the criteria of the scoring rubric.

Figure 4.6 Friend Conference Results Sample.

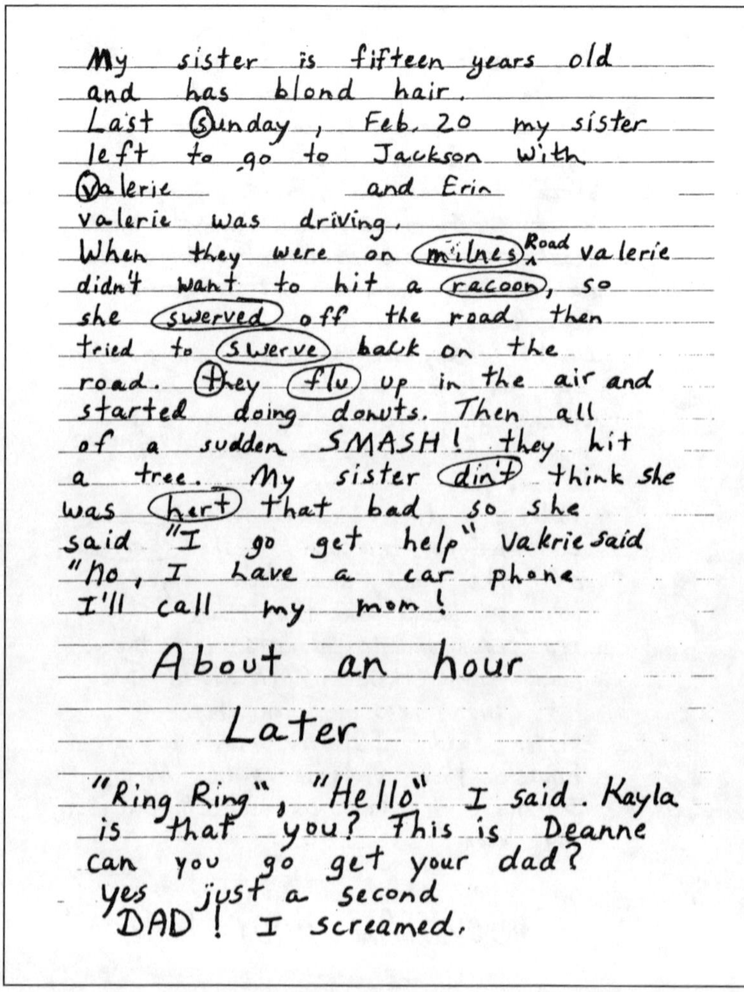
My sister is fifteen years old and has blond hair.
Last Sunday, Feb. 20 my sister left to go to Jackson with Valerie and Erin
valerie was driving.
When they were on milnes Road valerie didn't want to hit a racoon, so she swerved off the road then tried to swerve back on the road. they flu up in the air and started doing donuts. Then all of a sudden SMASH! they hit a tree. My sister din't think she was hert that bad so she said "I go get help" Vakrie said "no, I have a car phone I'll call my mom!

About an hour Later

"Ring Ring", "Hello" I said. Kayla is that you? This is Deanne can you go get your dad?
yes just a second
DAD! I screamed.

According to the state, a perfect score of "4" would show the following:

> Central ideas may be clearly developed through details and examples. The writing may have a natural flow and a clear sense of wholeness (beginning, middle, end); the organization helps move the reader through the text. A clear and engaging voice is likely to be demonstrated through precise word choice and varied sentence structure. Skillful use of writing conventions contributes to the writing's effect.

Although this is a holistic scoring rubric and the Lenawee County Portfolio Project fifth grade rating scale guide is analytical (with separate ratings for each category), there are similarities (development, organization, style and voice). Using the rating scale in my classroom helped me focus on some of the needs my students had as we approached the MEAP test in writing.

There are, in my estimation, some shortcomings of the test too. My students are allowed to brainstorm topics freely, where the state gives a topic. My students are allowed the freedom to publish pieces that they feel are "good enough" to publish, and ample time is given for this task. That is at odds with the three-day window allowed by the state. Lots of effort is expended and credit is given when my students revise and edit writing, where the state doesn't give opportunity for such effort nor opportunity for the credit it might deserve. Finally, I have a whole wonderful portfolio of student work to consider as I look at my student writers. I can see what we are doing well and what our needs are as I consider future lessons.

The state, unfortunately, sees one piece of work generated in a short period of time as the whole basis for its assessment, and, given the topic, it's quite possible that a proficient writer could have trouble generating a piece compatible with state requirements for a proficient score. It would be nice if the state could consider the whole body of student work and assess more as I do. This way, both the state and conscientious writing teachers could see the value of a portfolio rating scale as a way of meeting the needs of our writers. State review could become a place where authentic assessments hold value as we look at the whole student and make rational judgments about student learning.

Conclusion

So, where do we go from here? We see some important next steps in our development toward improving the quality of writing in our classrooms. We think that we need to continue to meet and further refine our lessons to continue to improve our students' writing. That means developing more effective lessons to help students in all the categories we have addressed. Another important step is to encourage other teachers to work with the rating scales. We need to find out what the other teachers are doing to help students with these categories of quality writing. Then we can better deliver a consistent program working with all the categories in the rating scales. We can imagine a school where all teachers are using a common language to discuss quality writing and using similar strategies to help develop quality at all grade levels.

Our school could have a tremendous writing program if we could consistently deliver instruction in these categories over a student's career. In order for this to happen we will have to take a strong leadership role. Fortunately, our schedule allows for grade level teams to meet regularly during the week. We need to make focused analysis of student work and improving our teaching practices a priority. The rating scales can help us with this focus. We also have a language arts study group that meets monthly before or after school, providing another opportunity for work with the rating scales.

We have known for several years that students in our district have been improving their writing and that they write very well. Now we have the opportunity to take our work to another level of quality. The rating scales are the framework from which we can help students achieve new heights as writers, and help teachers become more effective writing teachers.

CHAPTER 5

Two Teachers Help Student Writers Reflect on Their Portfolios: What Do They See in High School?

Kathryn Bell and *Liz Webb*

❖ ❖ ❖ ❖ ❖ ❖ ❖

As we moved toward best teaching practice in our high school classrooms, we struggled with assessment. We wanted assessment to align with the workshop approaches that we had implemented in our classrooms. Traditional grading practices didn't mesh with authentic assignments that focused on students' connections with their writing and reading. Having students read and respond to each others' work created a different emphasis and view of expertise in the classroom. The more we changed, the more we had to think how might these changes shape assessment practices?

Work Together to Figure Out Solutions

Fortunately, we worked together not only to identify problems but also to explore solutions. Many things made us a good team. We are two people who love good books, good conversation, and good teaching ideas. We looked for new learning experiences and new ways to improve our teaching practice so our students could blossom as readers and writers. We are also fortunate that the other high school English teachers and our K–12 language arts teacher-friends in our district are committed to workshop practices. Our district K–12 Language Arts curriculum is centered around reading and writing workshop practices and beliefs. Aligning our teaching practices and beliefs with compatible assessments has been an interesting, albeit at times angst-filled, process.

Seek the Help of Others

Finding a mentor who will share ideas and possibilities supports teacher growth. Dick Koch continues to be an invaluable resource to us as we learn more about workshop approaches and assessment. He's a trusted, knowledgeable guide, encouraging us to make changes and take risks in our quest to create the best teaching and learning possible.

As we were struggling with writing assessment in our third, week-long summer workshop class with Dick, he shared the Lenawee County Portfolio Project descriptors with us. We have never used them in exactly the same way they were used in Lenawee County, but they've been valuable as we continued to revise our practices and understandings.

Adapt Ideas to Your Own Situation

The ways we used the descriptors differed from the elementary practices described earlier in this book. When our small rural high school converted to a 4 x 4 block schedule, our entire district was also adopting reading and writing workshop approaches for our K–12 Language Arts curriculum. The high school English teachers spent a summer reconstructing our traditional year-long curriculum into genre-based quarter classes. The English 10 class, for example, became elective courses in poetry, expository writing, themes in literature, and biography.

With the block schedule, we don't work with our students for an entire year. Most of our classes last for only one quarter (nine weeks) or a semester (eighteen weeks). Instead of looking at student work written near the beginning, in the middle and at the end of a school year, we do so at the end of the quarter and the end of the semester.

Our goals for portfolio assessment at our level were also different. We wanted to find ways for students to be informed evaluators of their own work and the work of their peers. We wanted students to learn independently for a lifetime. Self-evaluation helped move them toward this school district goal.

Examine Traditional Practices

When we taught grade-level English courses (pre-block schedule), our courses resembled those in most high schools. Our seniors studied British Literature, did research papers, wrote literary analysis papers, and took tests. Many high school writers had a fundamental grasp of writing conventions and could usually get something down on paper in response to a class assignment. What they didn't seem to have was an accurate idea of how well their efforts succeeded. They didn't seem to know when or why they hadn't succeeded or have any enthusiasm for thinking about the process. What we saw in their papers wasn't necessarily what they saw, and having a conversation about those issues was sometimes difficult. We knew what constituted good writing and we were the sole evaluators of our students' work.

Share the Secrets

As experienced teachers, we could pretty rapidly read a student's piece and assign a letter grade. We used our insights as readers and years of experience to compare and assess. We didn't necessarily articulate or publish a specific set of criteria when we assigned these grades, however. When students wanted to know why they got the grade they did, we'd have to explain and justify our assessment to them. We'd go over their papers, pointing out stylistic, developmental, organizational and other issues that supported our evaluation.

This method had several flaws. Sometimes, we couldn't find justification through specific points in the paper although there was no doubt in our minds that the grade was fair and appropriate. Sometimes, students seemed to have little idea what we were talking about when we mentioned specifics that contributed (in our eyes) to the grades given. When those occasions arose, we felt as if we were drifting into the quagmire of "mystery learning and grading" that seemed depressingly like the landscape of much of our own high school experience.

This was one of the dilemmas of teaching that began to move us in new directions in teaching and assessing writing. We wanted to unravel the mystery of what constutes good writing with and for our students. We wanted them to know what we knew and we wanted to learn more ourselves.

A senior said it well when she wrote (after two years of immersion in the type of courses we now teach):

> It's very important for a writer to know his or her strengths or weakness in writing. Performing an in-depth self-evaluation does just that. It helps a writer see where they are strong, what techniques they are capable of, and where they need to grow. Often times young people are pushed to write but it goes no further. They may hand in a paper, get it back corrected, and toss it aside never to be picked up again. Out of sight, out of mind, right? With this mentality, one's writing can never become stronger, only serve as a means of getting an assignment done. Many don't understand that writing is much more than that. Writing is the key to other worlds, many of which are within ourselves (Deanna).

That's what we wanted our students to know and be able to do, but we didn't yet know how to help them accomplish this. Our unrealistic expectation, unspoken but deeply felt, was that students should be able, and willing, to examine their own writing efforts and compare them to the standards of the genre. Of course, wishful thinking never made anything happen educationally, and this was no exception.

As Donald Graves said in his essay "Help Students to Read Their Portfolios," in *Portfolio Portraits,* "For years we've assumed that students knew how to read their own work. I used to think, 'If they cared, they'd work harder and I'd have less to correct.' They do care. Once students become aware of what constitutes good writing, and get help with the elements that make it better, they work hard to make the quality of their writing match their original intentions" (p. 95).

As we explored this problem, we realized that we were expecting students to be able to self-evaluate when they hadn't been exposed to examples of what other students had tried to do with the same assignments. In traditional high school English classes, teachers give an assignment, students crank out one or two drafts in isolation and turn it in hoping they've somehow miraculously hit the mark the teacher had in mind. As Peter Elbow noted, "It's striking how often students actually say these words to you as they give you their papers, *I hope this is what you wanted*" (p. 219). When you're not sure what or where the target is, scoring a bull's eye is unlikely. This lack of student understanding bothered us, and we began to talk about what we could do to improve instruction and assessment.

Articulate Beliefs

As we discussed these problems of practice, we also started sharing our beliefs about learning. We knew we wanted students to be engaged, enthusiastic readers, writers, and thinkers, and we wanted to find ways to help make that happen. We discussed what we and others, especially researchers Brian Cambourne and Jan Turbill, think helps students learn well. These principles still guide our work with student writers. We think students need choices within the curricular guidelines that describe each course. If students are working on personal essays, for example, they can decide topic and treatment while we provide craft and genre advice. Choosing topics of personal value helps students engage with their writing in ways that didn't happen when we used to assign the topics. Thomas Armstrong notes, "Giving students choices is as much a fundamental principal of good teaching as it is a specific intrapersonal teaching strategy. Essentially, choice time consists of building in opportunities for students to make decisions about their learning experiences" (p. 83).

We also think students need to read what other writers generate as they work on the same assignments. Talking about writing with classmates and responding to those observations improves the writing our students produce. Because we believe student writers need time to write and respond, assignments often extend over several weeks. Sometimes students put pieces away to cool before returning to them with different ideas.

Encouraging students to set goals and expecting them to exercise judgments about their own and others' writing also involves them in their work in new and productive ways. When we began to ask students to reflect on their own processes and products, their understandings of what it means to be a writer and critic were reinforced and strengthened. Doing all of this with their classmates and teacher created a community of writers with norms, structures, procedures, goals and cohesiveness. This approach honors students as learners, rather than seeing them as producers passively dependent on the mysterious judgment of the more knowledgeable teacher/evaluator.

Once we started having these discussions and articulating our beliefs, we started looking for and creating ways to translate those beliefs into classroom practices. Finding ways to help students think about their own learning and efforts, as well as that of their classmates, was an important goal for us.

Move to Best Practices

Finding ways to engage our students in the process excited us because we'd experienced the power of the writing process personally, both with Dick and at the Traverse Bay Writing Workshop. We'd read the experts, including Nancie Atwell, Linda Rief, Regie Routman, Donald Graves, Donald Murray, Dan Kirby and Tom Liner, so we were well grounded in the philosophy of workshop approaches. But one of the basic things we still struggled with was how to structure opportunities for students to learn to self-assess. We knew in many instances that our students were capable of producing exciting pieces of writing, but they still needed lots of work to take their writing to publishable quality. If students couldn't self-assess, they revised superficially. How could we best facilitate this growth?

We wanted to move to portfolios because we were convinced that drafting, peer conferencing, revising and editing carefully, over time, would produce better writing from our students. Portfolios would allow all of us to view that total body of work.

We nearly drove ourselves and our students crazy, however, trying to devise a workable grading system. Initially, we were more interested in justifying the grades we placed on report cards than we were in creating ways for our students to evaluate their own growth in writing over time. We didn't realize then what valuable information we could glean from student response regarding classroom instruction and practice. We didn't yet understand how looking carefully at student writing could also help create a better English department and a better K–12 comprehensive Language Arts curriculum.

Adapt the Descriptors

When Dick shared the descriptors that the Lenawee County teachers had developed for their portfolio project, we felt we had found help. Our adaptation and use of the descriptors from that project helped us understand and work on all of the problems of practice we had been exploring. We added indicators to describe what our students might be seeing and doing as they developed in each of the categories. The indicators we created were useful ways for us to convey expectations to students about criteria we used to assess student writing (see Figure 5.1).

Adding the indicators let us articulate for our students those behaviors and criteria we used when we evaluated their work. Suddenly, it felt as if we were letting them in on the secret in extremely useful ways. We became partners in their development as writers and they now had a concrete, specific guide we could all share. Each year it becomes more apparent to us how valuable the criteria and practices that we adapted several summers ago have become for us and for our students.

Take the Plunge

We first moved in the direction of workshop approaches by adding primary research projects and directed independent reading to the English 12 courses

Figure 5.1 Lenawee County Descriptors with Specialized Indicators.

Descriptors	Indicators
Fluency is the ease with which a writer writes when presented with an occasion for writing, measured by the amount of writing in a given period of time or in a given situation. The goal is to generate and sustain writing more and more easily. Consistent fluency is one mark of a good writer who is in control of her or his writing.	The writer fills journal pages easily and uses pre-writing techniques effectively to get to an idea/draft. Drafts increase in length when it is appropriate and/or desirable.
Development refers to the amount of helpful information included in the writing, especially specific facts and/or sensory details. Appropriateness will depend on genre and rhetorical situations. Developmentally, at the first step, some facts and/or details will be included, although not all of these will relate to the subject of the piece. At the next higher level, facts and details will mainly relate to the subject, but some important information may be left out. At the highest level, nearly all the facts and details are related, and are relevant or essential to the subject and purpose of the piece. Nothing important is left out.	Content "shows" instead of just tells. Successive drafts answer reader questions. Pieces show awareness of audience.
Organization refers to the order of ingredients in a piece of writing and to the maintenance of focus on a central issue or idea. Organization also includes the use of effective transitions to connect the parts of a piece, which should be in a helpful order for the reader.	All parts of the piece deal with the topic of the piece. No unrelated or divergent ideas are included. The reader can easily follow the progress of the piece.
Style/Voice refers to the way the writer's personality comes through in the piece. It also refers to the word choices the writer makes, as well as the syntax and sentence variety in the piece. Developmentally, a writer first learns to handle these elements appropriately, then effectively, and finally with special imagination or power.	The piece sounds like a real person is talking. Words used are in the writer's repertoire. The writer selects words and sentence structure appropriate to reader and purpose.
Variety deals with how successfully the writer is able to deal with different writing genres, purposes, situations, and topics. Some writers focus on a single topic or type of writing for long periods because it is important to them at that stage in their development. All writers grow by trying various modes and topics.	The writer tries pieces about varying topics in different forms or modes (poetry, short story, essay, autobiographical narrative, persuasion).
Revision means returning to a previous draft and making it better in terms of the above areas of consideration. Revision makes use of feedback from others to reconsider a piece and make changes to it. It is also expected that a writer's repertoire of revision strategies will gradually increase. Also desired is an increasing awareness on the writer's part of which section or aspect of a piece most needs revision.	The length of pieces usually increases through several drafts. Successive drafts include actual visual evidence of change. Successive computer drafts are printed and changes noted. Development, syntax, diction and organization improve.
Editing is the final polishing that makes the piece publication perfect.	Correctness issues in the areas of mechanics (spelling, capitalization, punctuation, word usage, paragraphing) are attended to. This stage is also called proofreading.

we both taught. The primary research project asked students to select a topic of strong personal interest, locate mentors and craft learning experiences that allowed them to become exceptionally well informed in that area. At the end of the course, they created elaborate, engaging presentations to share their new knowledge and their learning experiences. Directed independent reading immersed students in choice reading, sharing, and study. Both additions to the traditional courses contained elements of writing workshop and we asked students to reflect on their processes and products. We had them begin keeping portfolio collections of their writing, including all drafts and conferencing feedback.

We focused on what we wanted to do with students and what we wanted to have them do as we continued to revise our practice. Luckily for us and our students, as we began studying and using writing workshop approaches in our high school classes, the writing descriptors and our indicators made this task easier. We know a good thing when we see it, and we also know how to massage, modify, experiment with, and apply what we think will improve student learning.

These descriptors helped students discuss specific qualities in their writing. They used them in many ways. Sometimes they self-assessed as an entry-level activity when they began a writing class and used their initial assessment to set goals for themselves as writers. More often students used them as self-reflection and assessment tools to examine their final portfolios. In an early effort at final assessment, we asked seniors to analyze their own writing by giving them the following instructions:

> ***Introduce*** this marking period's writing. Tell about each piece you produced.
>
> ***React*** to things you did in prewriting, to writing first drafts, to conferencing, to using the computers, to meeting deadlines.
>
> ***Discuss*** the role writing plays in your life.
>
> ***Identify*** the things that are difficult or frustrating for you in writing.
>
> ***Find, copy and explain*** your favorite sentences from this marking period's writings.
>
> ***Assess*** yourself as a writer in each category mentioned on the *writing assessment* sheet of descriptors and indicators. Support your self-assessment with examples from your writing this term and from the feedback you received from conferencing partners.
>
> ***State your writing goals for this course and assess your progress.*** What are your new writing goals?

We were asking them, in specific and intentional ways, to look at their own writing in ways that we and they hadn't done before. We found we were trying to help our students do what Donald Murray calls "teach yourself with commentary" by having them look at the feelings, techniques, solutions to problems, and process moves connected with their writing. Writing assignments became increasingly centered in students' own experiences. The focus in our classes was shifting, and we were asking students to become more active participants in their own learning and assessment.

Network with Others

As we discussed, created and wrote new curriculum, we were supported and influenced by our district's involvement with the Michigan Language Arts Frameworks (MELAF) project. As a demonstration site, we used the standards and best practice research generated and explored in MELAF sessions with other Michigan school districts to create new courses and instructional practices. Standards about student voice, meaning and communication, the genre and craft of language, and issues of quality helped us attend to these important areas as we created new courses. Each course contained elements of all the standards, as well as self-assessment and reflection by students. The Michigan Curriculum Framework content standard 12, Critical Standards, states, "All students will develop and apply personal, shared, and academic criteria for the enjoyment, appreciation, and evaluation of their own and other's oral, written, and visual texts" (p. 20).

Give Students Tools

If our students were to become effective evaluators of their own and others' writing, we needed to find ways for them to recognize, practice, and internalize elements of quality writing. We wanted them to leave us with the dispositions and understandings that would allow them to be in control of their own growth and development as writers and thinkers. We also wanted them to be able to analyze assigned and choice reading and their classmates' work with an eye for the elements that make writing strong and effective, as well as to be able to identify weaknesses in writing.

We clearly saw how the use of the descriptors could be extended for use by students themselves. Many books we read regarding alternative assessment mentioned metacognition again and again. We knew that finding ways for students to think about how and why they did what they did in the specific pieces of writing in their portfolios would deepen their understanding of what it means to write well. As Regie Routman noted in *Invitations,* "A portfolio approach—student centered and properly implemented—has potential for high-level self-evaluation. Engaging in self-regulation and self-examination of multiple data sources gives the learner the message, 'I am better at this today. Here is my evidence'." (p. 342). Our new courses provided the perfect opportunity for students to learn about and to evaluate their own writing processes and products.

Start with a Pilot Study

In our Advanced Reading and Writing class we crafted ways for students to examine their own and others' writing using specific criteria and definitions and made that an integral part of the course curriculum. From the first time this class was offered, students investigated and wrote in many genres and took their own efforts through multiple drafts. As students read and discussed their own and others' writings, they began to identify specific examples of quality and improvement in the pieces, as well as areas that needed still more work.

In addition to looking more closely at their own work, our students also began reading and critiquing the writing of their peers. As they noticed, suggested, and praised specific things about each other's writing, their own understandings grew.

> In writing workshop, I needed content conferencing. Was I answering all questions a reader might have? Was I showing rather than telling? Was my voice/style consistent? As the term progressed, I needed less help in all of these areas, as suggestions from other people made me more capable of fixing things by myself.
>
> I learned things from conferencing with other people, too. When I read many of Charlie's stories, I was struck by how report-like they sounded and suggested that he work on voice. When he did, the pieces sounded ten times better. This reminded me of the importance of strong voice, and I looked for that in my own writing. (Rachel)

Instead of us directing students' attention to such issues as voice, style, or tone, students were identifying these concerns for themselves through their use of the descriptors. Another valuable spin-off from this approach seemed to be a growing student awareness that a first try at a piece will usually not be adequate.

> I have learned that it takes many attempts at a piece before you get it exactly right. I had no idea that authors went through many drafts, and so many different revisions. This will be helpful in college, because I won't just do one draft and then hand it in. I will make several drafts of a paper so that the quality will be that much better. (Gregg)

Students who are being read by their peers seem to pay more attention to the quality of their writing than they used to when they were writing only for the teacher. They also begin to see which students have specific strengths as writers and conferencing partners.

> I think that everyone tried very hard at the beginning of the term to be good responders, and then slacked off a little. But by the time they slacked off, I'd already learned who to go to for what. For instance, Rachel is good at helping you with the flow and content of the piece, while Greg is very good at spelling, punctuation, and mechanics. He is also good at giving positive feedback about possible additions to the piece. (Carla)

When students learned these things about each other, they were able to solicit needed help from their peers. Students became aware of the many readers with expertise in our classes. In addition, students seemed more open to revision suggestions and began to notice more about their own drafts as they read the writing of others.

> I almost always see a huge amount of improvement when I revise a piece. I usually get the main ideas down in my first draft, but little else. During the later drafts, I go back and add detail everywhere I possibly can. I'm usually not very happy with my first draft of any piece; however I am confident that it will get better after conferencing and revision. (Peter)

When students were asked to assess their own writing using the descriptors, they often evaluated themselves based on the kind of reader feedback they received from their peers.

> In organization I gave myself a three because I started to go off on unrelated ideas. Some people who read my stories said they couldn't follow the story all the time so I had to work on that. Throughout the year I tried to work on transitions, too, so that the reader can follow the story line easier. (John)

Considering the needs of one's audience is an important step in the development of a writer.

Help Students Learn From Each Other

Because reading others' writing was such a powerful learning experience for our students, we created many sharing opportunities. Read-arounds, where students working on the same type of writing pass their pieces to others in the group for feedback, provide powerful illustrations of how different writers craft a genre in different ways. Considering the assignment, students could see models of what others had done well. Using the descriptors of good writing, they could point out areas of strength and suggest ways other areas could be improved. After a read-around, each student had a broader vision of possibilities and pitfalls and could look at his or her own writing through new eyes.

> When you read other authors' works, you can become inspired. You may think to yourself, 'I want to write just like this.' You can also gain ideas from other authors. They may use certain techniques that would benefit your writing, like foreshadowing, imagery, or humor. (Greg)

Another student agreed. His observations suggest that having writing goals helps student writers know what needs to be revised and reading each other's work gives suggestions for doing so.

> Writing workshop probably helped everyone work toward their writing goals. I feel the most effective aspect of writing workshop was the conferencing by peers. The read-arounds were very helpful, at least for me, when it came to making content changes as well as grammar and mechanical corrections. (Tim)

Another student's comments on the read-around process highlight the importance of feeling that the class has become a community of writers supportive of one another.

> The revision and editing processes seem to go more smoothly for me every time I do it. I used to hate to have to do revision and editing, especially on my own pieces, because it always seemed like so much work for so little improvement in my piece. Read-arounds were only fun because I had the chance to read everyone else's work. I find it exciting to get back useful suggestions and questions. I believe that it has something to do with the group of people I'm surrounded by. It seems like every time I read a suggestion or question on my paper, I can see how much better it will make my piece even before I do make the changes. Revision and editing no longer seem to be a pain in the neck. I see them as an opportunity to watch a cluster of words and ideas grow into an actual form of literary work. (Yvette)

The willingness to help one another generates positive engagement. The process itself actually creates enthusiasm in students who are working on revision and editing. When students view themselves as authors, they take an important step toward thinking critically about their writing.

Teach Intentionally

As our students engaged in conversations about their writings, we found ways to help them consider specific issues of quality. Examples of well-done sections of their own writings were put on overheads for the class to examine and discuss. They located and shared passages found in their independent readings. We asked them to attend to specific elements of good writing by this studying of models together. The on-going conversations invited the class to investigate important issues in writing.

In Advanced Reading and Writing, we crafted a whole-class exploration of humor writing, a difficult genre for most writers to do well. The study involved each student bringing in a sample he or she considered humorous and explaining to the class what made it funny. We listed characteristics and qualities they noted. We also looked over *Far Side* cartoons as students selected those they found funny. Trying to explain to their classmates a cartoon that required prior or esoteric knowledge helped them understand some of the assumptions writers (and cartoonists) have about their readers. They also started to realize that not everyone considers the same things humorous. We were ready to study together samples of humor writing by S. J. Perleman, James Thurber, Erma Bombeck, Joseph Heller, and Molly Ivins, and a discussion of humor by William Zinsser. These models helped students identify some of the techniques humor writers use, including absurdity, parody, satire, exaggeration, elevated language, sarcasm, understatement, overstatement, incongruity, allusion, and caricature.

Encourage Risk-Taking

As we studied these models, students tried writing three short humor pieces on place, dialogue, and situation. From their three conferenced first draft attempts, they each chose one to take through more revisions and submit for evaluation. The rubric criteria they developed included topic suitability and treatment, detail and development, effective beginning and ending, flow, wording and vocabulary, use of specific humor techniques, as well as mechanics and correctness. Their efforts ranged from genuinely funny writing to a good effort that missed the mark, but they all understood more about the requirements of the humor genre when we were done.

Peter's discussion of his struggle with satire in his piece about his former hometown, Ithaca, NY, shows the type of understandings students can arrive at after studying humor writing and taking risks.

> The most frustrating part of this piece was trying to include the right amount of sarcasm in it. I wanted to include enough to make fun of the city, but not too much, or I could give a bad impression. In my first attempt, I included too much sarcasm. It was funny, but the people who read it said that the description made them dislike Ithaca, even though they had never been there. This wasn't the reaction I wanted to get. So, for my second attempt, I took out a lot of the sarcasm, and tried to show that Ithaca was not all bad. However, this completely backfired on me. While the piece no longer gave Ithaca a bad reputation, it was no longer funny (which is pretty important when you're writing a humor piece). I finally got the balance right in my final attempt. This piece displays many of my writing talents, and it does it in a way that people can smile about. I want people to walk away happy that they chose to read my work. (Peter)

Not only does Peter's analysis show the effort he made to get just the right amount of sarcasm in his piece, it also reveals a strong sense of audience. Such concern for audience is another sign that a student has moved outside purely individual concerns about his writing to consider the relationship between a writer and his readers. When student writers make that step, it is safe to say they are engaged in the writing process.

After the whole class study of the humor genre, students separated into focus groups of their choice to study and produce writing in other genres that included memoir, poetry, feature writing, persuasion, college application essay, or personal essay. Together, they created rubrics for their genres as we had done for humor and evaluated their own and each other's drafts with that rubric.

They read each other's drafts many times through the process, helping to establish a community of writers. When they joined new focus groups, the skills they had been polishing as they defined quality within a genre and worked together to improve their pieces carried over into the new groups. At the end of the quarter, as students polished pieces for publication and evaluation, we asked them to document the processes and mental work they used while creating each piece.

Study the Completed Portfolio

The annotations we asked students to complete for each piece of writing became part of their course portfolio and gave us the individual stories behind each piece of writing. Students wrote the one to two-page annotations of the piece(s) they took to publication in response to the following questions from us.

- How did this piece begin? What prompt or inspiration led to this piece?
- What conferencing partners and/or techniques were particularly helpful? How? Why?
- What revision techniques worked? What did you change and why?
- What writing and/or correctness issues did you deal with as this piece evolved?
- What frustrations did you encounter? Why? How did you cope with them?
- Why did you choose this piece to take to publication quality?

The final thing we asked students to do was assess the growth and effectiveness of their writing. We asked them to use the descriptors and indicators to evaluate their writing and expertise in all eight areas. They used journal entries and drafts and the responses they received from classmates and us to help inform their assessment.

Use Student Reflection to Refine Instruction

We used this student feedback to evaluate and revise our own teaching and course content and help us determine what craft and skill issues needed more attention. In their final reflections, students were asked to answer questions that helped us determine what needed more emphasis. They told us what kinds of conferencing, response and instructions they liked or needed, and if this changed over time. We asked what correctness rules or issues they felt we should have spent more time on as they revised and polished their writing.

> I don't think we dealt enough with correctness issues as we revised and polished. Many times when I had questions about the grammar or punctuation in my piece, no one in class could help me; none of us knew how to deal with some of the more complex correctness issues. I think as a class we need to work more on when to apply commas and semicolons. We also need to work on staying in one tense; many people in class, myself included, have a nasty habit of switching from past tense to present tense inappropriately. (Rachel)

For Rachel, one-on-one conferencing about these issues and the references to the *Writer's Inc.* guidebook section on punctuation weren't enough for her to feel comfortable with commas and semi-colon use. For her, and others with the same concerns, more deliberate teaching of those skills could be added to the course. Rachel's feedback also told us that she understood what writing correctness issues she wanted to work on; that awareness is another important indicator of student development as a writer.

We asked students about the variety of writing they were able to try, and what helpful knowledge about writing they felt they'd acquired during the course. They told us which things they felt were important as they worked on their own writings or read someone else's work. We used that feedback to think about how we might even more effectively teach those skills.

Look at Growth Over Time

Because we teach courses elected by many of the same students, we were able to look at the work of shared students over a longer period of time. We tried a new type of self-assessment and reflection when we asked them to consider their work over a two-year span. They returned to their previous year's portfolio and self-evaluation and compared it to the work they had written and the understandings they had developed since then.

One student did even more than we asked her to do. She actually looked at four years' worth of writing. As Jennifer discussed the variety, style, and tone of her writing over four years, she found evidence of her significant growth as a writer.

> My style has changed over the past years. In ninth grade I was writing mediocre mysteries and the same is true for tenth grade. In eleventh grade I started to write with more variety, and my tone began to change. The style in which I wrote strayed from my traditional story crafting where one paragraph follows the other. In my favorite piece from last year, each paragraph possesses the ability to stand on its own; it does not need the rest of the work. Of course it all fits together, but that was a large step which I took. My short stories this year have been different from that. The story which I wrote for my creative writing class was not my style at all. It had very descriptive external elements surrounding the main character—I typically write about internal issues. Everyone who read it was shocked that I had written it. I was shocked; I did a good job, but I was shocked. What interests me so is that looking at this evolution laid in front of my eyes is like taking a step into my psyche from years ago. I can actually tell what stage of development, mentally, I am going through. (Jennifer)

In her reflection, Jennifer shows she possesses the ability to be a lifelong learner, able to notice significant evidence of her development and growth.

After practice, students often can trace exact achievements in their work, as Jennifer does in this explication of her poem.

> "I Hate Clicky Shoes," my second work, a creative poem, demonstrates good use of style and voice. The poem, I feel, carries my attitude towards

> the subject matter in a manner which is quite similar to my own voice. It definitely shows my love of words through the diction employed, using words like "quidnuncs." Another aspect of this poem that reflects style/voice is the way I play with the white space. The poem has short lines and words do not always fall where a traditional poet would place them. The placement of the tap, tap, taps throughout the work adds to their meaning—strengthens them, as I would strengthen them with my tongue. Slip and away are also placed in different patterns across the page adding to their subtle departure. Poetry is a spoken art, but with the right techniques, the muscularity of the poem can be conveyed across the page. (Jennifer)

Jennifer understands the concept of style and voice. Her reflection let us know that the intentional lessons already used in class to teach tone, muscularity, and the use of white space worked well for her. After reading a whole class set of papers and reflections, we could easily decide which concepts needed to be retaught, as well as what lessons on other aspects of style and voice needed to be developed.

Jennifer's reflection could also be useful for other student writers as a mini-lesson on the specific craft issues of style and voice. With permission from the writer, the teacher could display the poem and Jennifer's comments side-by-side on an overhead to teach these elements of good writing to other students, as well as to show other students what a quality reflection could contain.

Good student reflection discusses the criteria and quotes from the student's own work in ways that support the self-assessment. Students who use their reflections to set new writing goals demonstrate that they know what further refinements they need to make in their writing and that they recognize that becoming a better writer is an on-going process.

Let Writing Cool

Another technique that helps students look objectively and critically at their own work is "cooling time." Students who can leave a piece for a week or two and then return to it with an eye for revision possibilites are often more apt to find and use those opportunities. In the following reflection, Kirk articulates and understands the value of time lapse in distancing himself emotionally from the words and ideas contained in his own writing. We are all connected intimately to our words and sometimes struggle to look at papers objectively for purposes of revision and editing.

> I like the way this assignment is set up because it allows for a lapse of time between the time we wrote the piece and the time we reflect upon it. Trying to evaluate a piece shortly after finishing it is not always a good idea because you are still involved with the piece and are still thinking about it. Coming back to it a month or two later is a good idea because the paper can be looked at more objectively. After waiting a certain amount of time, it almost seems as if you are reading someone else's paper, allowing for better study and evaluation. (Kirk)

Students who internalize the process and are able to step back from their own writing are empowered to produce quality work.

Connect Writers and Readers

"Writing is simply the writer and the reader on opposite ends of a pencil; they should be as close together as that." Jay R. Gould (Safire, p. 53).

Students often produce analysis that reflects deep understandings. In the following reflection, Deanna started to discuss her organization and segued into a detailed account of development and fluency, but the complexity of the issues did not stop her. She was probably tougher on herself than her teachers would have been in the workshop setting. She knows what to do with future writing because she effectively used the descriptors to look critically at her own work.

On a scale of one to ten, I would give the area of organization in this essay an eight. I somehow got stuck in the "five paragraph essay" mentality and never matured enough in essay writing to venture out of the five paragraphs. Sometimes that format can get dull and limit your writing, which I think happened in this essay. If it's possible, I think it was a little too organized, or at least I restricted my ideas for the sake of organization, which is not always good. The biggest thing that I have learned in essay writing is the importance of development. In this particular essay I can definitely recognize the lack of specific examples and detailed and exact explanations. It seems that often I have a good idea but don't take it far enough so that the reader is just dropped off before the idea can have any impact. Really, what I created was the skeleton of an essay without the body, a mere summary.

Because of this lack of development, I believe the fluency of my essay was jeopardized. Rather than complete well thought-out explanations, the reader was given half-ideas where, just as they thought they were reading a good point, they were, as I mentioned earlier, dropped off. This experience can be compared to reading a bunch of fragmented sentences. It gets old fast! After reading a well written, fluent paper, the reader should come away with a feeling of settlement, like the writer has made his point. When a paper is not so fluent, as with this paper, the reader comes away exhausted, like he just swam for a while and got nowhere. The purpose is defeated. (Deanna)

Deanna's self-assessment is detailed and informed; she analyzed her writing against clear standards and decided where and why she found it well-done and what was lacking. Another important indicator of quality in Deanna's reflection is her awareness of the reader; writers who write with their readers in mind tend to more closely consider important elements of writing, such as development and flow. What Deanna saw when she looked at her own writing was no longer mysterious; she could understand where her work succeeded and where it still fell short, at least in her eyes.

Our teacher role began to shift when we realized that an important part of our task as teachers was to help students be able to do for themselves what we used to do by ourselves with their writing. As Katherine Anne Porter reminded us, "Writing is a craft. You have to take your apprenticeship in it like anything else" (Safire, p. 56).

These intentional approaches teach the building blocks of good writing to all students. They see and hear them over and over again. The air resonates with the words they need to eventually be able to discuss and use to improve their own writing.

Reap the Benefits

Self-reflection was especially helpful for students as they set goals for themselves. Their goals can become part of the summative evaluation at the end of the course. Have they reached their goals or not? What evidence can they cite from their writing to support their assessment? Students who set goals for their own learning realize greater success. Teaching them is more stimulating for their teachers, too.

As students became comfortable and competent discussing their writing, they began to engage creatively with the assignments we gave. Reading this kind of student work was so much more rewarding than when we'd laborously work our way through a class set of stilted and conventional essays on the same teacher-assigned topic. The more our students worked with revision and looked at their own writing, the better they became at both tasks.

Continue to Learn

As Regie Routman also noted, "The teacher-learner or student-learner (sh)ould be observing and talking about his own learning processes—including strategies being used, changes being noticed, progress being made, strengths and weaknesses observed, past and present levels of functioning noted, and new goals needing to be established" (p. 370). As we wrote about our learning and change in this chapter, we also talked about the new learning and change that we see coming in the future.

We wondered, for example, why and how some students made significant revisions and improvements to their writing when they engaged in the process while others did not. We found John Dixon's observation in *Growth Through English* comforting, "The teacher can do little more than make available a variety of situations in which, if the pupil chooses, the knowledge may gradually become explicit and controllable" (p. 73). Continuing to find more and better situations for students to engage with their own learning is a priority with us.

We also wonder how best to convince other teachers that this really is a better way to teach writing and encourage student growth when the demands of time and intellectual engagement are sometimes daunting. We made progress with student self-assessment, but we continued to wonder about ways to better communicate evidence of student growth to parents and others. We hope to always find ways to observe and study our own learning and evolution, and we will continue to construct opportunities for students to do so as well. In doing so, everyone benefits.

Grow as Writers, Grow as People

We believe what Howard Gardner said "What we need in America is for students to get more deeply interested in things, more involved in them, more engaged in wanting to know; to have projects they can get excited about and work on over long periods of time; to be stimulated to find things out on their own" (Graves, p. 146). Students who exhibit effort, engagement, and enthusiasm can excel in many areas.

When we found and developed better ways to get our students engaged in their own writing and to then help their classmates become better writers in the process, we felt we were finally realizing the vision we had of student learning and growth. Writing became a metaphor for the larger kinds of growth we wished for our students. Deanna said it well.

> Something I have come to understand is that revision and editing never end. I could keep this portfolio for the rest of my life and never stop adding new things and strengthening what I already have. Writing is a record of change. Some day I will look back at different accomplishments, different milestones and say, "That improvement in my writing improved me." (Deanna)

Chapter 6

Using Portfolio Criteria and Rating as a Basis for Helping Teachers Improve Their Practice: Professional Development

When I saw the way our portfolio rating system could help improve the writing of my first graders, I knew I wanted other teachers in my district to have access to this approach. What I didn't know was that within four years I would get the opportunity to make this happen, at least for other elementary teachers. Because of my involvement in writing assessment at the county level, and more specifically, the Intermediate School District's Portfolio Assessment Project, I was asked by the district to share my experience with portfolio analysis, and to develop a plan for gathering and assessing student writing at the elementary level. I was released from my teaching assignment a half of each day to conduct this work.

Once I was delivered the opportunity, this challenge seemed even more overwhelming. While I was excited at the prospect of helping others use portfolios to inform teaching and monitor student progress, I was anxious about where to begin. Given the approximately 100 elementary teachers in the district, and the limited number of hours I was released to conduct this work, I was concerned my efforts would become diluted, lacking purpose or meaning for all those involved. After some deliberation, and with helpful advice from my director, I focused my efforts toward one grade level. Third grade was chosen because it was the one grade level where the least amount of information regarding student achievement was available.

During the first year, all third grade teachers were released three times to conduct a formal assessment of student writing (November, February, May). I met with teachers at their individual building sites, to keep our working size between three and five people. Teachers were asked to bring a sample of writing for each student in their class to each meeting. This sample was to be representative of the student's best work for the previous eight weeks, and would include all copies/drafts of the piece. Teachers were encouraged to involve students in the selection process, and to inform them of the purpose of our work as we proceeded.

At our first meeting in early November, we began the day by brainstorming what we believed to be "elements of good writing." We compiled a detailed list and discussed specifically what those qualities "looked like" for students at the third grade level. Teachers agreed it was important for students to write longer pieces of text with greater fluency. They were concerned about detail, elaboration, and organization. The ability to "revisit" a piece was a priority, along with attention to editing and correctness issues. When we finished our conversation, I passed out copies of the third grade writing values and the definition of terms designed by the original Portfolio Research Team. It was clear there was a match between the discussion at hand and the work of the earlier research team. With minor changes, the materials and formats outlined in this guidebook were accepted and used enthusiastically.

At the beginning of the project many teachers had little experience using a scoring rubric. In order to get comfortable using this type of measure, and to ensure some consistency in rating, we practiced rating several pieces of student writing together. We first analyzed the writing of students from the files of those who participated in the Portfolio Research Project. Later, teachers shared pieces from their own classrooms. As with the selections I provided, pieces were rated openly, each teacher sharing their thoughts as they assigned a number. As the day progressed, the group showed greater consistency in ratings and rationale. They seemed ready to "dive into" their student's papers with focus and objectivity. At this point, teachers were given time to rate each student writing sample, on a scale from 1 to 5, in the areas of fluency, development, organization, style/voice, revision, and editing/correctness. A class composite was completed to determine class strengths and needs, and to help plan for writing instruction during the following weeks.

The winter and spring assessments followed a similar format. Each session began with discussion and feedback regarding the direction of our work. We read and rated several student papers as a group, each paper engaging us in new conversation about some aspect of writing. We talked about the power of the portfolio in helping us view student progress over time, and looked to the research for further understanding and direction. By the end of our first year, teachers seemed comfortable using a scoring rubric and were pleased with the amount of information they had acquired.

I continued to work with third grade teachers the second year. We again met three times during the school year for a fall, winter, and spring assessment. In the fall, we revisited the scoring rubric and rated some pieces together to refresh our thinking. I informed the group that this year I wished to devote a larger portion of time to specific elements of writing. Based on the feedback sheets collected from the district's professional development workshops, I planned mini-sessions that addressed the issues of revision, writing in the content area, and the reading/writing connection. I gathered articles, videotapes and handouts that were relevant to each discussion. Additional resources were provided for individual study outside our meetings. Each teacher received a copy of *The Art of Teaching Writing,* by Lucy Calkins, and Pat Cordeiro's *Whole Learning.*

At each session we discussed how to "teach for" those qualities we considered important to good writing. We shared how the writing process was implemented in our classrooms, and searched for ways to improve our daily instruction. On several occasions, I joined teachers for one-week visits during their classroom writing time. I shared my personal writing with students, modeled mini-lessons appropriate to classroom need, and conferenced with individuals regarding the direction of their pieces. Teachers frequently expressed concern with the process of revision. They felt their students were often satisfied with initial drafts and resisted any prompts that would require them to revisit their writing. They further commented that when students did make significant revisions, it often "made the piece worse." Writing became artificial, each piece resembling another. According to the teachers, "The writer's style and voice had vanished!"

For that reason, my work with students frequently began with mini-lessons that addressed this concern. Barry Lane, in his book *After THE END,* became one source that offered help as I prepared lessons that would "nudge" students in a new direction. In Chapter 5, Lane speaks of "exploding a moment." This involves taking a part or paragraph of a story and making it "come alive" for the reader. First, I found exploded moments in books that were familiar to students. Next, I shared how I used the method of exploding a moment to further develop a piece of my own writing. A story about a recent trip to San Francisco served as one example. I selected the following small part of the piece that I wished to expand upon:

> . . . we visited Monterey, Carmel and Big Sur, as well. The scenery was spectacular! We saw beautiful beaches with cliffs that looked like mountains. The hillsides were covered with colorful vegetation. We spent our days breathing in the beauty of the coastline.

Next I exploded the moment, with greater detail and elaboration:

> Perhaps my favorite scene was a secluded beach we found just outside of Big Sur. I remember the waves, deep and blue, smashing against the jagged shoreline. The cliffs seemed so majestic, never moving against the constant force of the water. The sand was a beautiful mixture of white and tan granules. It felt smooth and warm as it engulfed my feet. We walked the edge of the beach for what seemed like hours. I could taste the salt from the water as the cool breeze blew across my face.

As Lane suggests, we talked about the power of using our "binoculars" to expand upon a moment. Next we practiced "zooming in" with papers written by fellow classmates. Students seemed to connect to this form of revision and were eager to find those moments in their own pieces. I also compiled a packet of additional resources that would help teachers design their own lessons for revision. That packet would include:

- A bibliography of useful reading materials
- Several self-assessment forms for student reflection

- A sheet with personal thoughts from my study and experience

By the end of year two, I was encouraged by our efforts. It was clear that the time we spent analyzing student papers and planning for instruction based on these findings had significantly influenced our third graders' ability to generate, develop, and organize a quality piece of writing.

As a school district, collecting student work and rating that work against some established criteria provided valuable information about student writing achievement that could not be attained through a standardized measure or a "once a year" writing test. The tool allowed the district to view students as they worked to connect to certain writing techniques and then apply those techniques to their writing over an extended period of time. It also provided valuable feedback on instruction, informing teachers and administrators of "successes achieved" and "next steps" needed. The portfolio and the rating system became a foundation for a comprehensive assessment plan. This measurement approach was used to monitor growth in writing at the third grade level during the course of the school year. The following charts, Figures 6.1 through 6.6, show our record keeping of student writing progress in relation to each criteria: fluency, development, organization, style/voice, and editing/correctness. When interpreting the charts, keep in mind that a decreasing number of students on the left side of the graph indicates a fewer number of students at beginning levels, whereas an increasing number on the right side indicates more students achieving higher levels of ability.

As I entered the third year of my work, it seemed important to include a broader number of teachers and students in the portfolio process. However, time for study and discussion in a small group atmosphere remained essential. It would also be necessary to match my work with the planning currently in progress for the delivery of Language Arts instruction at the district and building level. Though each of the seven elementary schools adhered to the same Language Arts curriculum, the models

Figure 6.1 Writing Portfolio Fluency.

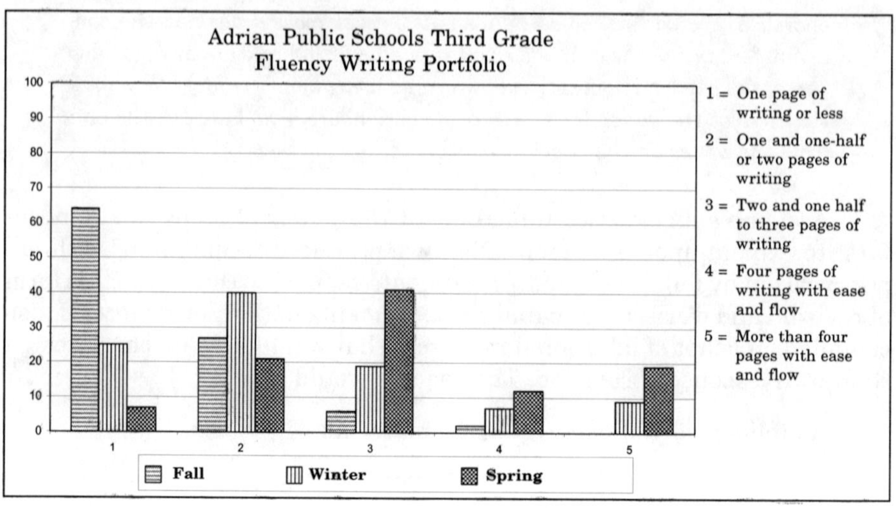

Figure 6.2 Writing Portfolio Development.

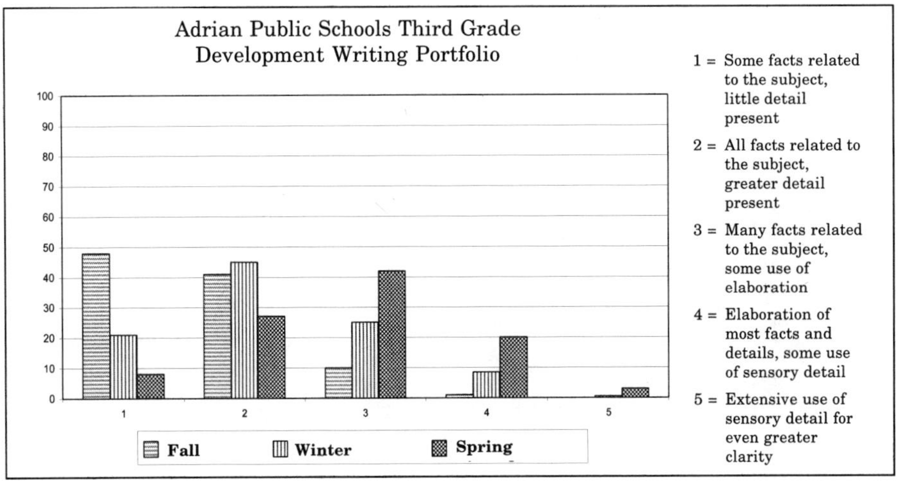

Figure 6.3 Writing Portfolio Organization.

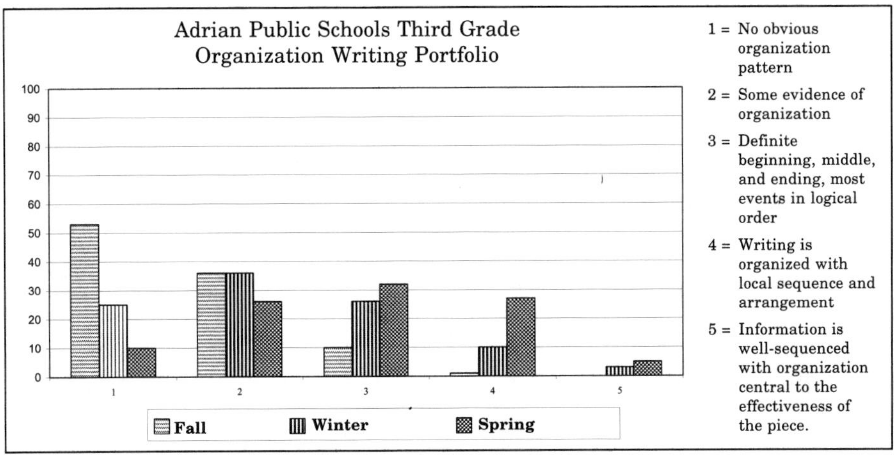

Figure 6.4 Writing Portfolio Style/Voice.

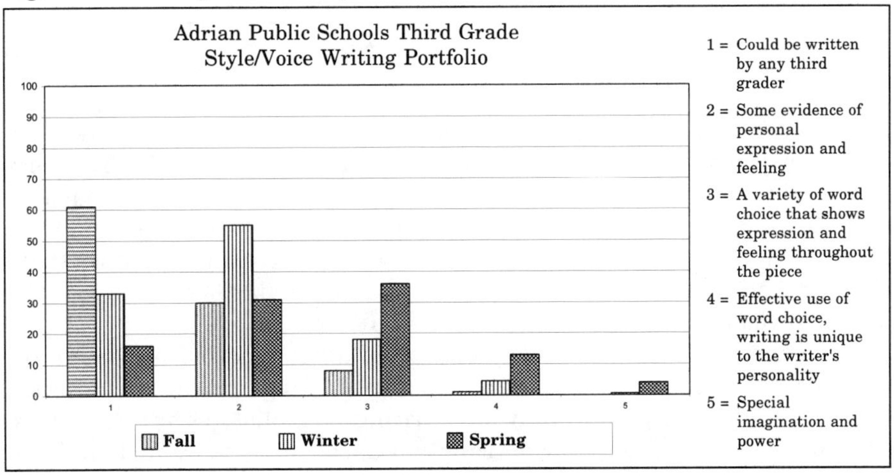

Figure 6.5 Writing Portfolio Revision.

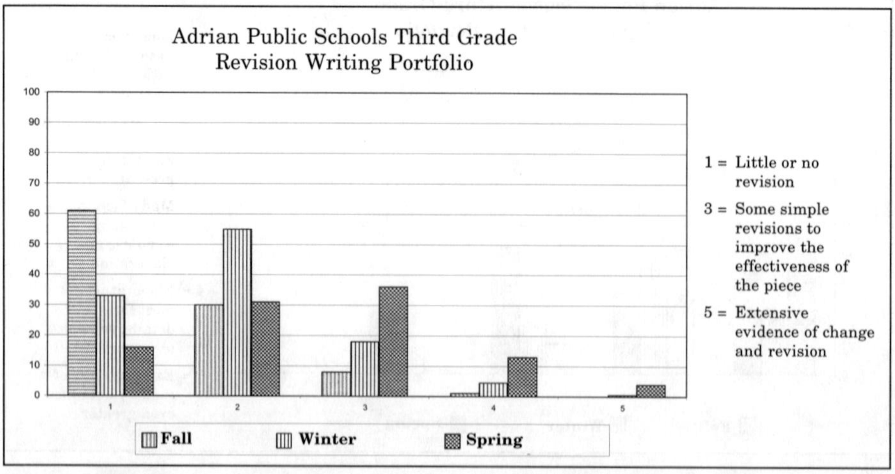

Figure 6.6 Writing Portfolio Editing.

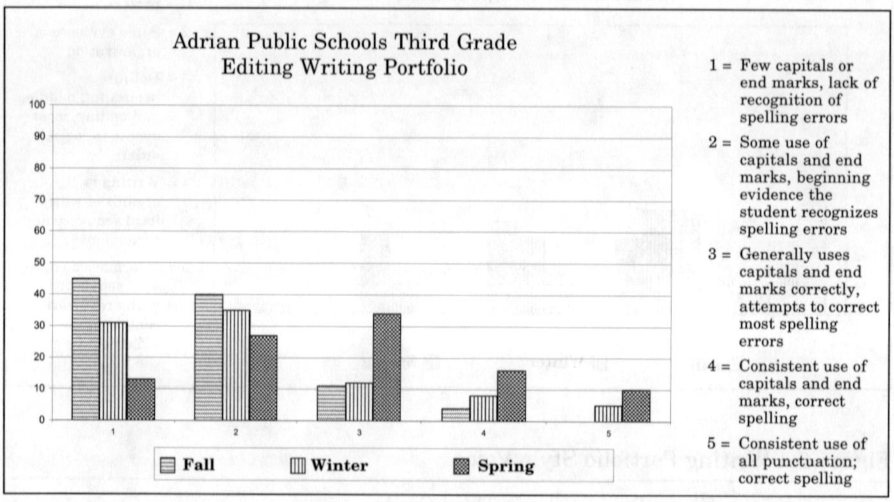

and methods used for implementation were different for each site. Two elementary buildings were involved with the *Learning Network.* Teachers in those buildings were working with the New Zealand model of teaching and learning, and were guided by teacher leaders who were released from their teaching assignments to assist with literacy instruction. Because those teachers had such substantive support, my efforts with these sites could be less involved than with those from other buildings.

To accommodate the needs of each building, I met with elementary administrators to plan for my work. Some buildings required continued support in gathering and analyzing writing portfolios. Other sites were now comfortable using a scoring rubric and looked for more instructional support. All sites were concerned with the state's writing test given at the fifth grade level, and looked for links between the district's initiatives and the state's expectations for elementary students. My year was devoted to providing support to teachers and students as we used the portfolio assessment tool to influence teaching and learning.

Collecting student work and analyzing that work on a regular basis seemed to produce even more useful and effective writing lessons. Teachers could easily identify those elements of writing that demanded attention, and could plan for instruction accordingly. Fourth and Fifth Grade Teacher Lori Faulkner conducts a formal assessment of student writing using the portfolio guidelines once each marking period. The piece chosen from the student writing portfolio is the one piece considered to be the student's best work for the previous eight weeks. The fall rating helps her establish "where the students are" and plan for writing goals the following weeks. Later, the rating system helps her notice "where students are going and what's new in their writing." For Faulkner, this assessment tool is "more specific and encompassing" than other measures. "It reminds me of what needs to be evaluated, so I'm not haphazardly leaving out some aspect of writing that is important and needs to be noticed."

Students are involved in this process as well. "We look at writing together so students can see what special areas of writing need improvement." The assessment model not only helps this teacher plan for better conferences, but also allows her students to "see their strengths and set some goals for writing the following weeks."

Third Grade Teacher Jeff Petterson shares similar thoughts. The rating system "gives me an opportunity to step back and see where my class has needs in terms of writing." He uses the assessment measure to conduct a fall, winter and spring rating. The tool is used to monitor student growth and plan for instruction more precisely. "In the fall, I recognized that while many of my students were fluent, (meaning they could generate a number of pages in writing) they weren't producing writing with interesting details or great elaboration. So the next month, I addressed that concern in my mini-lessons. It was a focus for whole class instruction." The same diagnostic approach was used for the winter rating. "I noticed that my students' work lacked 'personality or voice.' This became a topic for several whole class lessons the following weeks." Petterson also notes that the system for gathering and assessing student writing becomes quicker and easier with each rating. What seemed like "a lot of work" in the beginning, has now become an important and efficient use of time.

My work with writing instruction and portfolio analysis continues to change as we all take on new and deeper understandings of the role of writing in teaching and learning. A structured, district-wide system for gathering and assessing writing portfolios has now given way to individual building plans. The rating system has become a flexible tool used to meet the various needs of all those involved.

Currently, I provide support to the upper elementary grades as we begin to include more writing in Science, Social Studies, and Mathematics. Designing a tool that will monitor growth and inform teaching in these areas seems a logical "next step." Developing that measure will be challenging. A collegial effort will be necessary to construct a tool that will best notice those skills and achievements we deem important. I believe that a process similar to the one used to develop the materials for this guidebook will be valuable as we venture into that work.

I further believe that authentic assessment returns the content of instruction and the process of evaluation back to the classroom teacher. The countless hours currently devoted to preparing for the "yearly test" may instead be used to engage students in meaningful activities that build upon current levels of understanding. Authentic assessment helps diminish the "one size fits all curriculum," and works to narrow instruction toward appropriate need. The best teaching decisions can be made when the most accurate information regarding student achievement is available. Authentic assessment provides that information. It allows us to develop a more specific and in-depth profile of students and their abilities, leading to an educational experience that is more productive and promising for everyone.

As our students mature, so must all of us grow in our understandings of assessment and evaluation. The materials from this guidebook, and any further materials that may be developed, must be continually reviewed and revised to best meet the needs of the students and teachers it serves. No one method of assessment can be all-inclusive. However, we must consistently work toward the variety of assessments that will provide a more authentic view of the student.

When working with students at different schools I often ask the question, "How does your teacher know what to teach you in writing?" Students frequently respond, "My mom tells the teacher what to do when she comes to conferences," or, "My teacher looks in a book and it tells her what to do!" Students who have experience with portfolios, however, often have a different awareness. Daily tasks and assignments take on new meaning as these students attach more purpose to everyday activities. They not only benefit from instruction that is better informed, but come to understand that "my teacher decides what to teach me based upon what I have done, and what I need to learn next." In the words of one elementary student, ". . . he looks at my work and sees what I do, then he knows what I need to learn . . . he knows what good writing is like and he helps me to know that, too."

Chapter 7

Why This Type of Assessment Work is Essential to Achieving Quality—Conclusion

❖ ❖ ❖ ❖ ❖ ❖ ❖

When our original group of public school educators, joining with the county education agency and a local college teacher, agreed to begin a portfolio assessment experiment we did so because we felt a need to establish a closer tie between curriculum goals that we all believed in and assessment methods that would support those goals and even enrich them.

We felt that portfolios could be a key step in achieving this bond. It seemed to us that portfolio assessment was the most promising method available to help schools overcome the drawbacks of standardized tests and one-paper writing exams imposed from the outside.

It seemed to us that, with some extra work, it was actually quite simple and achievable to put our own better assessment into place. First, as explained in earlier chapters, we were aware that, if we began with a writing workshop classroom and if students kept a folder of their writings, carefully dating their work, students and teachers would have available a rich collection of writing. We reminded ourselves to include multiple drafts and a sequence of editing steps for certain pieces.

Second, we knew that if we agreed to select from this folder a representative sampling of the writing that would then form a "portfolio," we could set this aside to be examined more carefully, allowing both learner and teacher to view authentic writing performance and growth over time.

We hoped through our research to find positive answers to five important practical questions. If we had doubts they weren't about the basic worth of what we were doing, but rather about how fully and precisely we would be able to answer our questions. Those questions were:

- Can a group of teachers effectively describe what they want to see in student writing?
- Can they train themselves to precisely and reliably notice what they say they want to see?

- Does work with portfolios increase student learning and teacher understanding of what has been learned?
- Will portfolio assessment interact richly with curriculum goals and classroom methods to improve schools?
- Can portfolio assessment do all of these things without taking up too much teacher, student, and rater time, and without excessive expense?

After our study we feel we can offer a confident "yes" answer to all of these questions. Some of our answers came with significant data, and some are still only intuitions or glimmerings. However, even all of the glimmerings seem positive and clear.

We found, for example, that after four, three-hour training sessions, a staff of six raters were able to come to the point where they judged particular pieces of writing with reliable similarity—we *were* able to notice with exactness what we said we wanted to look for in the student writing.

This, of course, was only our own local confirmation of what Paul Diederich reported to the profession in the early seventies in *Measuring Growth in English*. Diederich had observed how a group of writing teachers who might begin wide apart in their ways and standards for viewing student writing could, through focused practice, become very reliable in their judgments.

We were able to confirm this, through triangulation in a variety of ways, including a "Blind Rating" experiment, after the regular rating session, in which all information, even the date, was removed from certain papers before they were read by a second team of raters. These results were consistent with those of the previous pair of raters who had been officially assigned to the grade-level group from which the special samples came.

We were also able to establish the clear possibility of using student portfolios for at least moderately large-scale assessment. In our case we were able to accommodate, with a staff of about six or eight teacher-researchers, and within a one-week rating period, work from early elementary, middle school, and high school. In this week we not only achieved reliable ratings but made up various final report charts to present different views of the student-writers' achievements to various potential audiences.

We believe that any school, or set of collaborating schools, that chooses to do so can very likely set aside time from professional development (this *is* professional development, after all), and/or from final exam week, or from some other relatively convenient period, to examine a fairly large body of student work. In our case, we set aside one week immediately after school got out, and we negotiated a modest but supportive stipend to be paid to the teacher-research team of raters for their professional time.

Not all grade levels would need to be assessed each year, and perhaps an "every-student-assessment" would not have to be the goal. Sampling is a time-honored, statistically legitimate method of analysis. And remember, at the point of large-scale assessment, we are talking about rating pieces of writing (not marking or responding to the writing), an activity that begins to go very quickly with practice.

As we indicated at the beginning of this book, we found that systematic portfolio assessment was not an "add-on" that we could afford not to have time for. Its benefits were too strategic to our central teaching and learning goals to allow ourselves not to engage in portfolio review. Why let a score from a test that involved no writing become the only basis for judging our students' language use?

Or, why let a narrow one-shot writing become the exclusive indicator of what they could achieve? Studies have shown that a one-sitting writing is a sure way to make students and teachers "look bad." Weak writers will write even less well than usual under the pressure and confusion of a testing situation. Sick or emotionally distracted writers will perform poorly. Some students will not respond with serious effort to the testing situation because it does not seem necessary to do so. Others will not be inspired or well-guided by the topic or prompt provided for the test writing. Others will simply perform less well under pressure as test-takers than they would have as writers. And, there are other reasons, related to cultural diversity or types of intelligences, for lowered results with one sitting.

We could see both the necessity and the definite possibility for avoiding what has been called the single snapshot view of our students and for achieving instead a view of them through a photo album of more authentic writings developed over time.

The most strategically useful data we generated in our study is reported in appendices to this book. In general, we found what we felt were six key benefits of portfolio assessment in writing, benefits that, to us, amounted to a command from our conscience to continue to make use of portfolio review. Portfolio assessment is needed because:

1. It allows professional educators to assess and evaluate according to agreed upon goals for students.
2. It allows educators to do their assessing of students through looking at authentic work over time.
3. Interpretation of results allows schools to consider effectiveness of particular teaching methods in their program, allowing for a more informed discussion of questions such as, "Are we helping our students work effectively with revision?"
4. It creates highly beneficial teacher collaboration, including fruitful dialogue about the relationship between assessment and curriculum goals.
5. It increases student-writers' awareness of their own growth over time, thereby also validly increasing their self-esteem. This awareness is achieved both because the teacher notices growth and comments on it, and because the students are actively involved by maintaining a folder of their work, helping to select pieces for review, and writing reflective assessments themselves.
6. It provides a solid foundation of information from which educators can begin to speak more forcefully and more credibly about what they and their students are achieving in schools.

We hope we have shown there are many reasons for moving in the direction of portfolio assessment. In the previous chapters we have seen the possibility for precise assessment of strengths and weaknesses of even beginning first grade writing. We have seen how teachers can use portfolio rating criteria on a daily basis for reflecting on the writing of particular students, noticing their differing abilities, and thereby making the most effective and helpful classroom decisions.

We have seen how two teachers can collaborate at the end of a marking period, or at the end of a year, to share their student work with one another and to revise their teaching by looking at the student work with the sharp focus provided by the lens of the rating criteria. And, we have seen how students can become able to reflect powerfully on their own writing with the use of clear portfolio criteria and teacher guidance on what makes for good reflecting.

The need for a move to such assessment is also more pressing today than ever before. For some time now the best and most informed teaching—Whole Language/Integrated Language Arts, Cooperative Group Work, Reflective Learning, and the more recent national move to "Standards-Based Teaching"—has progressively made formerly relied upon standardized measures even more obsolete. If we do not take steps such as changing to portfolio assessment soon we will increasingly find ourselves in a preposterous *Alice in Wonderland* world where teachers and students may be doing improved teaching and learning but invalid assessments do not reveal what is being learned.

Teachers and administrators often feel that the demands and the pressures of their work are such that there is not time for "one more thing." We can readily sympathize with and relate to this feeling from the vantage point of our own work. However, very often the "one more thing" most neglected is the integrated authentic assessment that is actually essential if we hope to make sense of the "whole thing."

In an age of increased pressure on educators to demonstrate achievement through student results it is important that educators participate strongly in defining what those desirable results should be and how they will be measured. We no longer live in a time when other stakeholders in the country are willing to simply accept a teacher's, or a school's, word that "our students are doing well." State and national standards for quality are in existence or being written for all of the basic subjects (language arts, social studies, math, and science). In an age of public suspicion toward schools and of increasingly complex descriptions of standards for student achievement, it is crucial that educators become able to talk about quality and make sense.

The truth of the past is that often we educators have not collaborated well enough with each other on assessment to enable us to make sense in our explanations of quality from one grade level to the next. Particularly in regard to writing and reading, students, parents and others might often hear differing goals and criteria from one teacher to another.

Portfolio review of the sort we've been describing can be a giant step toward putting collaborating teachers on the same page with one another on the question of quality, and such review can allow teachers and students to notice and describe quality achievement in such a way as to allow for better decision-making on both their parts.

In a helpful book, *Evaluating Literacy*, the authors argue at some length a point we present here briefly—that schools should develop a sufficient variety of rich assessment methods so that all of the stakeholders in the educational experience are accommodated with helpful communications.

We would suggest further that by long tradition in this country the stakeholder most neglected in evaluation and assessment has actually been the student. Typical grading practice has been to repetitiously point out how far short of perfect student rote memory is at any given time ("You still have three words wrong on your spelling test"). The replacement, in some cases, of the so-called objective test with a one-sitting writing opportunity, we have argued, is also a grossly inadequate step.

And yet, such state-required or district-mandated tests and one-time writings have profoundly influenced school practice, reducing much of a student's school experience to drill and memorization in order to, supposedly, perform well on such tests. Instead of working over time on relatively more authentic reading and writing tasks in relation to more real-world standards of quality, the students practice narrowly for the testing moment. The damage this has done to student learning, and to our national level of literacy, is incalculable.

We shudder to estimate the number of classrooms still existing in this country that would reduce their year-end observation of our first grader Michelle's literacy to a standardized test score, or that would offer a final judgment of her writing that simply made note of the number of words that remain spelled incorrectly. Happily, Michelle and her writing received far more productive responses.

We educators must take our advice from the assessment revolution currently going on at the research level and assertively move new productive assessments, such as portfolio review, more solidly into the classroom. Until we do this the biggest stakeholders in American education, Michelle and her fellow students, will continue to be at risk of reductive evaluation responses and the subsequent damage to their potential as learners and human beings.

References

Anthony, R., et al. (1991). *Evaluating literacy*. Portsmouth, NH: Heinemann.

Armstrong, T. (1994). *Multiple intelligences in the classroom*. Alexandria, VA: Association for Supervision and Curriculum Development.

Atwell, N. (1987, 1988). *In the middle: Writing, reading and learning with adolescents*. Portsmouth, NH: Heinemann.

Braddock, R., Lloyd-Jones, R., & Schoer, L. (1963). *Research in written composition*. Champaign, IL: National Council of Teachers of English.

Calkins, L. (1994). *The art of teaching writing*. Portsmouth, NH: Heinemann.

Cambourne, B., and Turbill, J. (1988). *The whole story*. Auckland, New Zealand: Ashton/Scholastic Limited.

Cordeiro, P. (1992). *Whole learning and content in the upper elementary grades*. Katonah, NY: Richard C. Owen.

Darling-Hammond, L., and others. (1995). *Authentic assessment in action: Studies of schools and students at work*. New York: Teacher's College Press.

Diederich, P. (1972). *Measuring growth in English*. Urbana, IL: National Council of Teachers of English.

Dixon, J. (1970). *Growth through English*. Champaign, IL: National Council of Teachers of English.

Elbow. P. (1973). *Writing without teachers*. New York: Oxford University Press.

Fletcher, R. (1996). *A writer's notebook*. Portsmouth, NH: Heinemann.

Fletcher, R. (1993). *What a writer needs*. Portsmouth, NH: Heinemann.

Gardner, H. (1992). Multi-Genre Research: One College Senior, *Portfolio Portraits* (Graves and Sunstein). Portsmouth, NH: Heinemann.

Glazer, S. M. & Brown, C. S. (1993). *Portfolios and beyond*. Norwood, MA: Christopher-Gordon.

Graves, D., and Sunstein, B. (1993). *Portfolio portraits*. Portsmouth, NH: Heinemann.

Harwayne, S. (1992). *Lasting impressions: Weaving literature into writing workshop*. Portsmouth, NH: Heinemann.

Lane, B. (1993). *After THE END*. Portsmouth, NH: Heinemann.

Michigan curriculum framework. (1996). Lansing, MI: Michigan Department of Education.

Murray, D. (1996). *Crafting a life in essay, story, poem*. Portsmouth, NH: Boynton/Cook.

Murray, D. (1990). *Write to learn*. Chicago, IL: Holt, Rinehart and Winston.

Perrone, V. (1991). *Expanding student assessment*. Alexandria, VA: Association for Supervision and Curriculum Development.

Rief, L. (1993). *Seeking diversity*. Portsmouth, NH: Heinemann.

Routman, R. (1994). *Invitations: Changing as teachers and learners*. Portsmouth, NH: Heinemann.

Smith, F. (1981). Myths of Writing, *Language Arts,* 58, 792–798.

Smith, M. A. & Ylvisaker, M., editors. (1993). *Writing portfolios*. Berkeley, CA: National Writing Project.

Tierney, R., Carter, M. A., & Desai, L. E. (1991). *Portfolios in the reading and writing classroom*. Norwood, MA: Christopher-Gordon.

Turbill, J. (1985). *Now, we want to write*. Rozelle, N.S.W., Australia: Primary English Teaching Association.

Writers INC: A guide to writing, thinking, & learning. (1989). Burlington, WI: Write Source.

Zinsser, W. (1990). *On writing well*. New York: Harper Collins.

Appendix A

Rating Criteria Rubrics for Grades 1–12

❖ ❖ ❖ ❖ ❖ ❖ ❖

Appendix A includes the essential materials for readers to implement the rating ideas described in this book. The "steps" and "tools" you need in order to gather portfolios and rate papers as we did are all included in Appendix A. In the following pages you will find: "Teacher Checklist" for gathering portfolios, "Rater Instructions," "Plans for Summer Assessment Process," our "Terms and Definitions" for what we view as important qualities in good writing, the "Writing Portfolio Assessment Form" that represents our analytical scale—this is the form we used to rate papers, and "grade level rating values" for every level—grades 1–12.

Teacher Checklist for Gathering Student Writing Portfolios

1. Invite Dick Koch to class on the day when portfolios will be selected from the year-long student writing folders.
2. Set aside class time to do the portfolio selection.
3. Ask students to select their two pieces of writing for the portfolio and to write "Writer's Choice" clearly on the top of the first page of each piece they select.
4. Then, the teacher circulates and, possibly with student assistance, selects three pieces (including all steps of each one, any multiple drafts, or any steps from free writing to more finished writing). The teacher then labels each piece as representing "Early," "Middle," or "Late," in the student's writing development. Note: The late piece should represent the peak of the student's writing achievement, according to the teacher's best judgement. This piece should be from the last few months of the school year, but it need not be the last piece written by the student.

If some of the teacher-selected pieces are the same as the student selections, the pieces are simply labeled twice at the top to indicate their "dual role" in the portfolio.

5. The teacher and students double-check to see that all pieces are dated and arranged properly.
6. The students write a letter of introduction to their portfolios including the following ideas:
   - Why the student selected the pieces she/he chose, and what she/he likes about the writings.
   - What the student hopes a reader will notice about the different writings in the portfolio.
   - How the student feels she/he has grown or developed as a writer this year, and her/his view of strengths and weaknesses at this point.
7. The teacher writes a letter of introduction to the portfolios, which clarifies, in brief:
   - What contexts or assignments the writings come from.
   - What her/his writing goals for the students were (early, middle, and late in the year)
   - How she/he views a high and low portfolio according to the Assessment Rating Form.
8. All of these materials are gathered together in a box, and the box is labeled according to the school, teacher, and class.
9. The teacher labels and adds Affective Survey's from early and late in the year to the box.

Writing Portfolio Rater Instructions

1. Select two pieces to do two separate ratings on per folder—a piece that represents the student's starting point and a piece that represents the student's fullest change from that starting point. The second rater rates the same pieces selected by the first rater.

2. Each rating sheet has student level and I.D. number on it. Rater puts her/his rating code letter on each rating sheet completed.

3. Rate 10 and compare ratings. Write preliminary guide for using the definitions of terms at the grade level in question.

4. Tag exceptional improvements, great moments, and quotable material.

5. Rate remainder and compare. Try to reconcile all differences of more than one. Create a third joint rating sheet to represent compromise.

6. Write a second version of rating sheet definitions.

7. Review definitions as a whole group.

8. A third rater rates all papers with a difference of more than one in any two ratings.

9. Write a small collection of insights, guesses, and advice for next year (insights about the papers and the rating process, advice for during the year and the next summer rating process).

10. Write a positive comment and a suggestion for the teacher.

Plans for Summer Assessment Process

1. Assessing to occur over three, five-hour days (9–2).
2. At four portfolios per hour, 28 portfolios can be read and assessed in seven hours.
3. Daily Schedule:

Day 1

9–12	Practice rating sessions in whole group, refinement of rating techniques, last minute questions about the process, working toward final reliability standard.
12–1	Lunch
1–2	Begin rating process

Day 2

9–9:30	Troubleshoot any problems that came up the first day.
9:30–12	Rating
12–1	Rating

Day 3

9–12	Rating
12–1	Lunch
1–2	Write "rater explanations" of what a "1" a "3" and a "5" are at the grade level the rater is working with and for each category on the Rating Form.

Writing: Terms and Definitions

Fluency

Fluency refers to "ease with which the writer writes when presented with an occasion for writing."

- So, fluency has to do purely with "amount" of writing in a given period of time, or in a given situation. We would like students to be able to generate and sustain writing more and more readily throughout their school career.
- Therefore, given a certain amount of time for journal writing, or free writing, over a period of weeks or months, we wish for the student's typical journal-write or free-write to grow longer and longer.
- Ultimately, we have no desire to encourage teachers to press a student past a "peak fluency" point, nor do we wish, overall, to press so hard for fluency, or to value it so much, that we drive students regularly toward nonsense as opposed to "thinking through writing." However, consistent fluency is one mark of a good writer who is in control of his or her writing process.

Development

Development refers to the amount of helpful information included in the piece of writing.

- Development has to do not just with number of words but with the amount of specifics (facts and/or sensory details) included. And, unlike fluency, development also has a clearly "qualitative" aspect to it.
- The amount that is desirable will vary for different genres and situations.
- Although volume is an issue, "appropriateness" to genre and rhetorical situation also are issues relevant to successful development.
- There are three basic steps in quality:
    1. The lowest quality that is still somewhat successful (at least as a practice step) is the inclusion of some facts and/or details, or an increasing number of these, but with the problem being that not all of these facts and details are related to the subject of the piece of writing.
    2. The facts and details are, for the most part, "related" to the subject of the paper, but some important information is left out.
    3. Nearly all of the facts and details are not only "related" but either "relevant" or "essential" to the subject and purpose of the piece of writing. Nothing important is left out.
- So, both volume and relevance are important aspects of development. However, volume is a valuable practice step that will later allow for the teaching of the issue of relevance.

Organization

Organization refers to the "order" of ingredients in the piece of writing and to the maintenance or focus on a central issue and related issues.

- Ordinarily, one main issue should be present in a piece, and the order of elements should clarify and enhance our understanding of that issue.
- Some obvious signs of organizational control, or lack or it, would be the way the writer handles an attempt to deliver chronology, and whether, in an explanatory or persuasive piece, the introductory material comes early and the concluding material comes late.
  - We do not want to encourage any additional emphasis in the schools on "lockstep" organizational patterns ("always start with a thesis sentence," etc.), however, we want to assess whether parts are in a helpful or hindering order. In other words, we would assess whether clarity and meaning are enhanced or harmed by the order of parts of the piece.
  - Organization can be an issue at the level of the whole piece and at the level of the paragraph.

Style/Voice

This category refers to the use of voice (personality) and word choice in the writing.

It also refers to the quality of control of sentence structure and sentence variety (but not correctness of sentences).

We look at this area of the writing first in terms of whether those issues are handled appropriately, then whether they are handled effectively, and finally whether they (sentence voice, and word choice) are handled with special imagination or power.

Variety

Variety refers simply to whether the writer demonstrates ability to deal successfully with different writing situations and topics. This is less essential and less reliably measured than the three categories listed above.

- If a writer is successful in the areas of fluency, development, and organization over time, that writer will likely have demonstrated success with "variety" also.
- Over time, we might wish for the upper level student to be presented a variety of opportunities for writing.
- However, since it is probably appropriate for primary students to be involved most extensively with free writing, journal writing, and learning log and observation writing, we do not want to think that these students have "failed" to demonstrate variety.

- Furthermore, apparently obsessive treatment of a particular topic by a writer is not necessarily harmful (even famous writers sometimes deal with a topic obsessively). So, we want to notice variety (teacher-offered or self-selected) over time, but we want to be very discriminating and restrained in our observation of variety.

Revision

Revision issues also will be different for different age students. In general, revision means returning to a previous draft or free writing and making it better in terms of the above areas of consideration on our rating form.

- At the primary level we are looking for the beginning inclinations in the writer to make use of feedback from others to reconsider their piece and make changes in it, and for a student's repertoire of revision strategies to gradually increase.
- We look for these developments keeping in mind the tendency at certain ages to feel "all done" as soon as a preliminary draft is completed (we do not expect a teacher to spend an excessive amount of time pressing for revisions, when it would probably be more productive for the student to be encouraged to work toward fluency and development).
- We also keep in mind that not all pieces (and in some situations, only a small number of pieces) should be revised.
- At the upper level we also look for signs of use of feedback and for an expanding repertoire of revision strategies.
- In addition, we can look for, and hope for, an increased precision and appropriateness of revisions—what is revised is the part, or aspect, that most needs revising. And, after one or more attempts, the revision is successful.

Editing

Editing refers to "correctness" issues. As indicated on the Assessment Form itself, these issues are fundamentally different from primary level to upper level.

- The degree of emphasis on this issue should be modest through elementary school and not excessive at the upper level.
- We look for correctness to improve gradually in student writing, and we look for signs of appropriate teacher help with correctness issues. Appropriate help would occur in the context of writing and through mini-lesson or workshop steps, with correctness concerns applied at the final, or "editing" stage of writing.

Writing Portfolio Assessment Form					
(effectiveness) (amount used)	**ineffective not used 1**	**low little 2**	**moderate moderate 3**	**good often 4**	**excellent always 5**
Fluency					
Development					
Organization					
Style/Voice					
Revision					
Editing (Mechanics: spelling, capitalization, punctuation, word usage; Elementary: picture, lines and letters, invented spelling, correct)					

First Grade Rating Values

Name ______________________ Fall ❑ Winter ❑ Spring ❑

Fluency

1. A picture
2. Signs of print awareness (copying a word or attempting to write a thought)
3. Three or more thoughts
4. Up to a page of thoughts
5. Extended pages

Development

1. A message through picture
2. One piece of information or thought (a label)
3. Three pieces of information related to a topic
4. Provides more than three pieces of information related to the topic
5. Exceptional use of facts and details

Organization

1. Message is through picture
2. Organization mostly through labeling
3. Little sense of story, limited to sequence or statement of a few facts
4. Opening sentence, details and a conclusion
5. Same as above but with definite signs that the writer consciously attempted organization

Style/Voice

1. Style and voice through picture
2. Basic information, little description, minimal signs of personality, fragments
3. Attempts sentences consistently and uses some descriptive language to show personal involvement
4. Variety in word choice; writes in sentences using descriptive language appropriate for topic
5. Includes criteria above, but with greater sophistication

Editing/Correctness

1. No evidence of correct spelling, punctuation, capitalization
2. Attempts correct spelling, punctuation, capitalization
3. Evidence of internalized sense of correct spelling, punctuation, and capitalization appropriate to age level

Second Grade Rating Values

Name ______________________ Fall ❑ Winter ❑ Spring ❑

Fluency

1. Less than one page
2. One page
3. Two pages
4. Three pages
5. Four or more pages

Development

1. Few facts; not related to the subject of the piece
2. Most facts related to the piece; no supporting details
3. All facts related to the piece; some supporting details
4. All facts related to the piece; most facts are supported with details
5. Exceptional use of fact and detail—the piece includes elaboration and sensory experiences

Organization

1. No obvious organizational patterns
2. Some evidence of organization; beginning, middle *or* ending
3. A beginning, middle *or* ending; a logical progression of ideas
4. A beginning, middle *or* ending; most facts and details in logical sequence
5. Highly organized—the student arranges sentences with a particular organization in mind

Style/Voice

1. Lacks a sense of "sentence"
2. Sentences present; little description and minimal expression of feeling
3. Complete sentences; some use of descriptive language or feeling
4. Shows variety in sentence structure and word choice; uses descriptive language and feeling
5. Effective use of sentence, word choice, and expressive language

Revision

1. Little or no revision present
2. Some revisions to improve the quality of the piece
3. Extensive revisions to change and improve the piece

Editing/Correctness

1. Few capitals or end marks; lack of recognition of spelling errors
2. Greater use of capitals and end marks; shows evidence of recognizing spelling errors
3. Uses capitals and end marks correctly; attempts to correct most spelling errors
4. Uses capitals and end marks correctly; corrects spelling errors
5. Consistent use of capitals and end marks, and other punctuation; correct spelling

Third Grade Rating Values

Name ____________________ Fall ❑ Winter ❑ Spring ❑

Fluency

1. One page or less
2. One and one-half or two pages
3. Two and one-half to three and one-half pages
4. Four pages or more with ease and flow to writing
5. More than four pages with ease and flow throughout

Development

1. Some facts related to subject; little detail present
2. All facts related to subject with some detail added
3. Many facts and details with some elaboration present
4. Elaboration of most facts and details; use of some sensory detail
5. Extensive use of sensory detail for even greater clarity

Organization

1. No obvious organizational patterns
2. Some evidence of organized pattern (a beginning, middle, and/or end)
3. Definite beginning, middle, and ending; most events in logical order
4. Writing is organized—logical sequence and arrangement
5. Information is very well sequenced with organization central to the effectiveness of the piece

Style/Voice

I. Could be written by any third grade student
2. Some evidence of personal expression and feeling
3. A variety of word choice that shows expression and feeling throughout the piece
4. Effective use of word choice; writing is unique to the writer's personality
5. Special imagination and power in word choice and/or phrasing

Revision

1. Little or no revision present
2. Some simple revisions to improve effectiveness of writing
3. Extensive evidence of revision to change and improve piece

Editing/Correctness

1. Few capitals or end marks; lack of recognition of spelling errors
2. Some use of capitals and end marks; beginning evidence that the student recognizes spelling errors
3. Generally uses capitals and end marks correctly; attempts to correct spelling errors
4. Consistent use of capitals and end marks; correct spelling
5. Consistent use of capitalization, end marks, and other punctuation; correct spelling

Fourth Grade Rating Values

Name ______________________ Fall ❑ Winter ❑ Spring ❑

Fluency

1. One page or less
2. One and one-half or two pages
3. Two and one-half to three and one-half pages
4. Four pages or more with ease and flow throughout
5. More than four pages with ease and flow throughout

Development

1. Facts related to the piece; little detail present
2. Facts with details related to the subject of the piece
3. Facts and details related to the piece; some elaboration present
4. Elaboration of facts and details throughout the piece; greater use of elaboration present
5. Extensive use of elaboration of facts and details; nothing important is left out

Organization

1. No evidence of any organizational patterns
2. A beginning, middle, or ending; the piece rambles
3. A beginning, middle, and ending; some information is out of order
4. A beginning, middle and ending; all events and facts are logically sequenced
5. Information is very well sequenced with organization central to the effectiveness of the piece

Style/Voice

1. Could be written by any fourth grade student
2. Slight evidence of personal expression and feeling
3. Moderate evidence of voice and personality
4. Effective word choice; writing is unique to the writer's personality
5. Extensive variety in sentences; feeling and expression throughout

Revision

1. Little or no revision present
2. Some simple revisions to improve the quality of the piece
3. Extensive evidence of revision to change and improve the quality of the piece

Editing/Correctness

1. Few capitals or end marks; lack of recognition of spelling errors
2. Some use of capitals and end marks; beginning evidence that the student recognizes spelling errors
3. Generally uses capitals and end marks correctly; attempts to correct spelling errors
4. Consistent use of capitals and end marks, correct spelling, some paragraphing
5. Consistent use of capitals, end marks and other punctuation; correct spelling and paragraphing

Fifth Grade Rating Values

Name ______________________ Fall ❑ Winter ❑ Spring ❑

Fluency

1. One page or less
2. One and one-half or two pages
3. Two and one-half to three and one-half pages
4. Four pages with ease and flow
5. More than four pages of print, exceptional ease and flow

Development

1. Facts related to the piece, little detail present
2. Facts with details related to the subject of the piece
3. Elaboration of facts and details throughout most of the piece
4. Facts, details and strong collaboration of the two throughout the entire piece
5. Extensive use of facts, description and detail; nothing important is left out

Organization

1. No evidence of an organizational pattern
2. A beginning, middle, or ending; the piece rambles
3. A beginning, middle, or ending; information is generally in order
4. A beginning, middle, or ending; all events and facts in logical order; the story is easy to follow
5. Very well ordered; the piece is solid and complete; the writer uses a definite organizational pattern that aids the effectiveness of the piece

Style/Voice

1. Could be written by any fifth grade student
2. Slight evidence of personal expression and feeling
3. Moderate evidence of voice and personality
4. Effective word choice; writing is unique to the writer's personality
5. Special imaginative power in words and phrasing; feeling and expression throughout the piece

Revision

1. Little or no revision present
2. Some simple revisions to *improve* the quality of the piece
3. Extensive evidence or change and revision to *improve* quality of the piece

Editing/Correctness

1. Lack of punctuation; extensive use of invented spelling; poor mechanics
2. Some invented spelling; correct punctuation throughout *most* of the piece; mechanically sound
3. Correct spelling throughout *most* of the piece; uses correct punctuation; proper mechanics
4. Correct spelling; correct punctuation; attempts to use exclamation points and quotation mark where appropriate; attempts correct paragraphing
5. Complete correctness in every way; virtually no mistakes

Sixth Grade Rating Values

Name ________________________ Fall ❑ Winter ❑ Spring ❑

Fluency

1. One page or less
2. One and one-half or two pages
3. Two and one-half to three and one-half pages
4. Four pages of print with ease and flow
5. More than four pages of print with exceptional ease and flow

Development

1. Mostly facts; few details
2. A few details; some adjectives are unrelated or misused
3. Details are appropriate; could include more
4. Appropriate facts and details throughout
5. Extensive detail, nothing important is left out

Organization

1. Rambling, not well organized
2. Hint of beginning, middle, end; some events are out of order
3. Definite beginning, middle, end; logical or helpful sequence
4. Easily read and followed; effective sequencing
5. Well ordered, definite sense of pattern, complete, solid; organization increases the effectiveness of the piece

Style/Voice

1. Simple words and sentences; voice seems out of control
2. Varied some hint of personality
3. Person's voice and some precision seen throughout
4. Voice adds to mood and helps draw in the reader
5. Extensive style and voice; special imagination; powerful

Revision

1. Ineffective, or worse than original, or no evidence of revision
2. A few changes to increase clarity
3. Moderate revising that improves quality somewhat
4. Purposeful, effective changes
5. Strong, thoughtful changes that significantly improve quality

Editing/Correctness

1. Lack of mechanics; extensive invented spelling
2. Some invented spelling, caps on proper names and beginning of sentences
3. Most spelling correct, sentences end; punctuation correct; capitalization correct; appropriate paragraphing
4. Accurate spelling, few usage errors, correct caps and end marks, some commas; appropriate paragraphing
5. Correct spelling, usage, commas, end marks, understanding of quotes; virtually error free

Seventh Grade Rating Values

Name ______________________ Fall ❑ Winter ❑ Spring ❑

Fluency

1. One page or less
2. One and one half to two and one half pages
3. Three pages
4. Four pages
5. Four or more pages

Development

1. Only basic facts, almost no details
2. A few descriptive details added to extend facts
3. Colorful verbs and multiple descriptive devices used in limited amount
4. Details promote reader involvement, fairly consistent throughout
5. Extensive detail, all relevant, nothing important is left out

Organization

1. Disjointed, rambling
2. Loosely connected beginning, middle, end
3. Definite beginning, middle, end; logical or helpful sequence
4. Easily read and followed; effective sequencing
5. Solidly organized, clear intent, organization increases effectiveness of piece

Style/Voice

1. Simple words and sentences, voice seems out of control
2. Varied some, hint of personality
3. Person's voice and some precision seen throughout
4. Voice adds to mood, helps draw in reader
5. Extensive style and voice; special imagination; powerful

Revision

1. Ineffective or worse than original or, no evidence of revision
2. A few changes to increase clarity
3. Moderate revising that improves quality somewhat
4. Purposeful, effective changes
5. Strong, thoughtful changes that significantly improve quality

Editing/Correctness

1. Lack of mechanics
2. Some invented spelling, caps on proper names and beginning of sentences, appropriate paragraphing
3. Most spelling correct, sentences end, punctuation correct, capitalization correct
4. Accurate spelling, few usage errors, correct caps and end marks, some commas
5. Correct spelling, usage, commas, end marks, understanding of quotes, virtually error free

Eighth Grade Rating Values

Name ________________________ Fall ❑ Winter ❑ Spring ❑

Fluency

1. One page or less, struggles with words
2. Two or three pages, little struggle with words
3. Three to four pages, comfort with expression
4. Four to five pages, words flow smoothly
5. Five or more pages, ease of expression

Development

1. Facts, but little or no details
2. Facts and attempts a few details and or reasons; some not appropriate
3. Facts and several appropriate details and reasons expressed as single adjectives or words not part of a greater unit of meaning
4. Appropriate facts and details and reasons, developed in some unified sections; elaboration of ideas throughout
5. Extensive facts, details and reasons throughout, all relevant

Organization

1. Little or no plan
2. Simple follows natural order of the topic (e.g. first we did this . . . then we did that, or in first grade . . . in third grade)
3. Some sense of control of sequence
4. Controlled sense of sequence
5. Controlled, purposeful sense of order and sequence; order increases effectiveness

Style/Voice

1. Little or no evidence of writer's personality
2. Some evidence of the writer's personality found
3. Evidence of the writer's personality often found, some precision of language
4. Voice adds to mood, evidence of the writer's personality throughout
5. Controlled presence of the writer's personality found throughout, special imagination, molding voice clearly toward audience

Revision

1. Ineffective or worse than original, or, no evidence of revision
2. A few changes to increase clarity
3. Moderate revising that improves quality somewhat
4. Purposeful, effective changes
5. Strong, thoughtful changes that significantly improves quality

Editing/Correctness

1. Lack of mechanics, extensive invented spelling
2. Some invented spelling, caps on proper names, and beginning of sentences, appropriate paragraphing
3. Most spelling correct, sentences end, punctuation correct, capitalization correct
4. Accurate spelling, few usage errors, correct caps and end marks, some commas
5. Correct spelling, usage, commas, end marks, understanding of quotes, virtually error free

Ninth Grade Rating Values

Name ______________________ Fall ❑ Winter ❑ Spring ❑

Fluency

1. One page or less, struggles with words
2. Two or three pages, little struggle with words
3. Three to four pages, comfort with expression
4. Four to five pages, words flow smoothly
5. Five or more pages, ease of expression

Development

1. Facts but little or no details
2. Facts and attempts a few details and or reasons; some not appropriate
3. Facts and several appropriate details and reasons expressed as single adjectives or words not part of a greater unit of meaning
4. Appropriate facts and details and reasons, developed in some unified sections; elaboration of ideas throughout
5. Extensive facts, details and reasons throughout, all relevant

Organization

1. Little or no plan
2. Simply follows natural order of the topic (e.g. first we did this . . . then we did that, or in first grade . . . in third grade)
3. Some sense of control of sequence
4. Controlled sense of sequence
5. Controlled, purposeful sense of order and sequence; order increases effectiveness

Style/Voice

1. Little or no evidence of writer's personality
2. Some evidence of the writer's personality found
3. Evidence of the writer's personality often found, some precision of language
4. Voice adds to mood, evidence of the writer's personality throughout
5. Controlled presence of the writer's personality found throughout, special imagination, molding voice clearly toward audience

Revision

1. Ineffective change; no evidence of change; slight change
2. A few changes, perhaps in wording and phrasing
3. Attempts to polish word choice and sentence structure, and adding detail
4. Polish and adding parts and significant content, a sense of recasting structure with a definite improvement
5. Significant global changes (e.g. rethinking the piece or substantial, effective rearrangements of parts)

Editing/Correctness

1. Multiple errors, ignores conventions, no sense of sentence structure
2. Many errors, little or no sense of paragraphing
3. Occasional errors in conventions
4. A few errors in minor conventions
5. Essentially error free

Tenth Grade Rating Values

Name ______________________ Fall ❑ Winter ❑ Spring ❑

Fluency

1. One page or less, struggles with words
2. Two or three pages, little struggle with words
3. Three to four pages, comfort with expression
4. Four to five pages with ease and flow
5. Five or more pages, smooth flow and confidence

Development

1. Generalities with no detail or examples
2. Some specifics with sketchy examples
3. Detail present in examples
4. Examples with clear sensory detail, as well as helpful facts or reasons
5. Developed images and complete scenes portrayed; array of facts and reasons developed

Organization

1. Skips around with no apparent plan
2. Organized around events or ideas as they seem to have occurred to the writer with little sense of structure for purpose
3. Has a plan with a definite beginning, middle and end
4. Writer is in control of plan and has a definite sense of its direction and sequence
5. Writer has arranged material in some original way and the idea shows a sense of effective movement

Style/Voice

1. Without individuality or sense of the writer's presence, could be written by any student
2. The writer asserts her presence with an occasional phrase or choice of appropriate word
3. A sense of the writer's presence is in the piece with regular use of words, phrases and sentence structure appropriate to the topic
4. A definite sense of the writer in the piece with frequent use of words, phrases and sentence structure appropriate to the topic
5. Writer shows control of style/voice through effective manipulation of sentence structure, word choice or other device (e.g. multiple voice, tone, etc.) and connection with her audience

Revision

1. Ineffective change; no evidence of change; slight change
2. A few changes, perhaps in wording and phrasing
3. Attempts to polish word choice and sentence structure, and adding detail
4. Polish and adding parts and significant content, a sense of recasting structure with a definite improvement
5. Significant global changes (e.g. rethinking the piece or substantial, effective rearrangement of parts)

Editing/Correctness

1. Multiple errors, ignores conventions, no sense of sentence structure
2. Many errors, little or no sense of paragraphing
3. Occasional errors in conventions
4. A few errors in minor conventions
5. Essentially error free

Eleventh Grade Rating Values

Name ____________________________ Fall ❑ Winter ❑ Spring ❑

Fluency

1. One page or less, struggles with words
2. Two or three pages, little struggle with words
3. Three to four pages; comfort with expression
4. Four to five pages with ease and flow
5. Five or more pages; smooth flow and confidence

Development

1. Unrelated facts or generalities with no detail or examples
2. Some specifics with sketchy examples
3. Detail present in examples and some fairly full images
4. Examples with clear sensory detail, as well as facts or reasons
5. Developed images and complete scenes portrayed; arrays of facts and reasons developed

Organization

1. Organized around events or ideas as they occurred to the writer with no sense of structure or purpose
2. Evidence of a plan that is not consistently followed, paragraph structure inconsistent
3. Has a plan with a definite beginning, middle and end, paragraph structure consistently appropriate
4. Writer is in control of plan and has a definite sense of its direction and sequence
5. Writer has arranged material in some original way and the idea shows a sense of effective movement

Style/Voice

1. Without individuality or sense of the writer's presence, could be written by any student
2. The writer asserts presence with an occasional phrase or choice of appropriate word
3. A sense of the writer's presence is in the piece with regular use of words, phrases and sentence structure appropriate to the topic
4. A definite sense of the writer in the piece with frequent use of words, phrases and sentence structure appropriate to the topic
5. Writer shows control of style/voice through effective manipulation of sentence structure, word choice or other device (e.g. multiple voice, tone, etc.) and connection with her audience

Revision

1. Ineffective change; no evidence of revision; or slight revision
2. A few changes, perhaps in wording and phrasing
3. Attempts to polish word choice and sentence structure and adds detail
4. Polish and adding parts and significant content, a sense of recasting structure with a definite improvement
5. Significant global changes (e.g. rethinking the piece or rearrangement of parts)

Editing/Correctness

1. Numerous sentence errors (e.g. fragments and run-ons, little evidence of correct sentence structure with many usage errors)
2. Many errors, little or no sense of paragraphing
3. Occasional errors in conventions
4. A few errors in minor conventions
5. Essentially error free

Twelfth Grade Rating Values

Name ____________________ Fall ❑ Winter ❑ Spring ❑

Fluency

1. One page or less, struggles with words
2. Two or three pages, little struggle with words
3. Three to four pages, comfort with expression
4. Four or five pages with ease expression and flow
5. Five or more pages, smooth flow and confidence

Development

1. Unrelated facts, or generalities with no details or examples
2. Most facts related to topic but lacks detail or explanation
3. Inclusion of facts and details that are related to the topic that makes the piece a unified whole
4. Shows the ability to extend and clarify ideas in a unified piece, all parts relevant
5. Ideas fully explained with effective array of facts, reasons, or details which create unity in the piece

Organization

1. Limited organization
2. Somewhat organized, may lack introduction and/or conclusion
3. Ideas presented in a logical or helpful sequential order and focus is maintained
4. Evidence of planned organizational development as it relates to the piece of writing
5. Obvious thoughtful, planned control of the piece of writing, relating form to content and purpose

Style/Voice

1. Little evidence of voice (presence of the writer), little variety in sentence or word choice
2. Some evidence of voice, appropriate word choice and sentence structure
3. Appropriate use of sentence, voice, and word choice
4. Evidence of writer's voice is present and obvious attempt at controlling word choice and sentence structure, mostly effective
5. Special imagination and power concerning use of sentence, voice, and word choice, figures of speech, etc., with effective audience awareness

Revision

1. Ineffective changes; no evidence of change; slight revision
2. Limited attempts at revision that improve the piece somewhat
3. Obvious attempts at revision, some areas improved
4. Revision significantly improves piece
5. Rethinking the piece, imaginative changes for the better

Editing/Correctness

1. No evidence
2. Limited control of editing
3. Occasional errors in conventions
4. Few editing errors in minor conventions
5. Essentially error free

Appendix B

Abstract of Report on Portfolio Assessment Project—First Year Results

❖ ❖ ❖ ❖ ❖ ❖ ❖

Appendix B contains key elements of our actual "first year report" from our original three-year-study of early elementary, middle, and high school students. It opens with our "letter to area educators," provides a short version of key aspects of the work such as "design of the study," "rater reliability," and "results" at the three grade levels. We also include the "Amount of Change" worksheets we used to calculate and report preliminary results. Only the first grade worksheets are included, although we used the same sheets for calculating middle and high school results.

To: Selected Lenawee County Educators
From: The Portfolio Assessment Research Project Team
Subject: Results of the First Year of our Three Year Study

Dear Fellow Educator,

Enclosed is your copy of the Assessment Report Abstract which clarifies results from each level in our first year of studying student writing through Portfolio Analysis.

As the report indicates, we studied two sections of tenth grade, one section of sixth grade, and two different classes (at the same school) of first grade.

The results are exciting! We feel we have established that even first grade level students and teachers can maintain folders and select from those folders to set up a portfolio, and this can be done without excessive time or difficulty.

Further, we have found ourselves able to achieve "rater reliability" with only four, three-hour training sessions spread throughout the school year.

And, finally, and most excitingly, we see at every level of our study significant growth in student writing—growth that could not be nearly so well observed by any other assessment method.

It is certainly our key recommendation that educators who would like, as soon as they are willing to, begin to make use of the relatively new but extremely promising assessment tool.

A key benefit of portfolio assessment is that, to quote from our report, "it lays a solid foundation of information from which educators can begin to speak more forcefully and credibly about what students are achieving in school."

We will be glad to assist school districts, individual schools, or any teachers who wish to join us as a "practice site" in our study so that they too can move toward use of portfolio assessment of student writing and reading in their schools and classrooms. There will be an Assessment Awareness Session at the Lenawee ISD from 4:00–5:30 on October 21 for those interested in this possibility. Please call if you are interested in attending.

First Year Results

A little over one year ago a collaborating group of Lenawee County educators decided to try to ease the pressure of standardized tests and to work toward better control over their curriculum development and evaluation processes by agreeing to participate in a three-year study of student writing development.

The study employed a portfolio assessment approach designed to examine samples of student writing that were generated early, middle and late in the school year.

Our study began with the involvement of three different levels of students: first graders, sixth graders, and tenth graders, with the different grades coming from three different school systems in the county. One of our goals is to follow a core group of students over three years at each school.

Our three-year plan also is to move toward an approximately equal focus on writing and reading in our study. However, for various reasons, including ease in getting our research started, we focused on writing during this first year.

Portfolio Assessment

The portfolio assessment approach may be the most promising opportunity available for schools to take hold of the assessment process in reading and writing, and thereby to actually measure student progress in areas that coincide with each school's curriculum and goals.

If we begin with a process or writing workshop classroom, and if students keep a folder of their writings, carefully dating the writings, the students and teachers then have available a rich collection of writing, including multiple drafts and a sequence of editing steps for certain pieces.

From this folder, students and teacher can select a representative sampling of the writing that can then form a portfolio to be examined more carefully, allowing both teacher and learner to view growth over time.

Design and Results of The Study

Design

It was important to us from the start to work toward practical, usable tools that would not require excessive amounts of class time or overtime. We wanted everything we did to ultimately provide us with teacher-friendly methods, limited school costs, and an increase in student learning rather than a distraction from it.

The following terms, rated on a five-point scale, represent the traits the Assessment Staff settled on as their guide to what to look for in student writing.

Fluency—amount written, ease with which the writer writes
Development—ability to include material which extends and clarifies, reasons, facts, or details
Organization—arrangements of material is helpful and contributes to effectiveness
Style & Voice—there is a sense of the person speaking (as appropriate to purpose), effective control of sentences
Revision—ability to improve the piece through successive drafts
Editing and Correctness—appropriate level of control of spelling, punctuation, and usage

These terms represent qualities that we feel are basic to good writing, and our definitions and guidelines have allowed us to apply the terms successfully.

Results

Results of a study such as this one must be judged, ultimately, on two basic criteria:

1. To what degree the rating staff were able to function accurately in relation to their own rating system

2. To what degree the ratings seem to actually shed light on student writing growth over time

We feel we succeeded in relation to both of these criteria, but we also feel that our work can benefit from a great deal of refinement.

Rater Reliability

In terms of reliability, our ratings show that the raters agree closely on achievement of individual student papers. If we consider ratings that are within one number of each other to represent basic agreement, then our raters achieved virtually complete agreement. However, we implemented two steps in our system to ensure reliability—a " reconciliation" or agreement of the two raters, based on a short meeting over a particular paper, when their ratings differed by one number, and an "adjustment" that was made by subjecting the paper to a third rater when the rating difference was more than one number. The reconciliations were infrequent enough, and the adjustments rare enough, so that we have already achieved the reliability needed to do this work efficiently on a yearly basis.

So, our amount and style of training are probably sufficient for our use, especially when you consider that a rating team may well improve their reliability in future years with practice, additional training, and an even more refined understanding of their goals and methods.

Our training approach, with four, three-hour practice sessions worked well. However, it is the feeling among all in our team that less training would not have been adequate. A "Blind Rating" experiment was conducted, in which a new rating team was given papers they had never seen before that had the dates removed and then were asked to use the rating criteria and guidelines for that grade level. This study, though it included only a small sampling, seemed to support our sense that we had achieved rater reliability.

Results on Student Growth by Grade Level

High School

In the high school results the decreasing numbers of students at the low end of ratings, and subsequent increasing numbers of students at the higher end of the ratings over the course of the year, showed a heartening sense of writing progress.

For example, there may have been as many as eleven students who were at the next to lowest level in development early in the year, and that number was reduced to six students at that level by the end of the year. And, there were only two students at the highest level in development early, and that number had increased to six students at that top level by late in the year.

The growth at the high school level in style & voice was especially impressive. According to our ratings, there were no students still remaining at the lower end of the scale (at the "1" or "2" level) in that category by late in the year. Furthermore, there was an increase from 10 students at the next-to-highest level ("4") early in the year to 19 students achieving at that level by late in the year.

Middle School

The middle school portfolio review was the most difficult in the first year of our experiment. The heart of the difficulty came at the point of gathering the portfolios. A flawed step in our process caused us to end up with fewer writing workshop pieces than we intended to review.

Because we often found ourselves comparing a personal narrative writing workshop piece from early in the year with either a research paper or a science report from later, we felt our rating in fluency and development were probably not particularly useful, even though those results, on the basis of numbers of the ratings, suggested progress.

However, the improvement indicated by students moving up to higher ratings in the other categories deserves to be noted and celebrated. For example, the number of students at the weakest level in organization went from five students early in the year to only one student later. And, similarly, there were six students at the lowest level in Style & Voice early in the year and no students in that low category later. We also observed a few students who moved into the second highest category in organization and into the second highest or highest category in editing by late in the year.

First Grade

In our study it appeared that first graders given regular opportunities to participate in writing workshop flourished and developed like perhaps no other age group. And, we ultimately felt this improvement could be observed and analyzed much better through portfolio assessment than by any other method.

The two first grade classes both made dramatic progress in writing. Our ratings suggest at least three impressive growth steps.

First, in regard to both fluency and development, we start the year with most students in the bottom two categories (at the "1" or "2" level). By the end of the year there are no students left in either class at the lowest level in fluency and only three students overall left at that level in development. In fact, by late in the year the majority of students in one class were in one of the top two categories ("4" or "5") in both fluency and development.

The third outstanding achievement appears to be in editing, where an overwhelming majority in these classes started at the lowest level, which is certainly what you would expect of first graders who are probably just beginning to write. By the late point in the year there were no students at the lowest level in one class and about half as many as had begun at that level in the other class in editing.

The group that made somewhat less progress in editing, however, made the most progress in organization, experiencing a large-scale movement of students from the lowest level early all the way to the middle level by late in the year. Both classes improved substantially in all of these mentioned categories.

Perhaps the most important progress in first grade is in the area of fluency, since ease in writing leads directly to confidence in writing.

Changes We Plan

First on our list of things to improve upon is our intention to manage the portfolio gathering process even more carefully, so that in every case the raters will have all of the materials they need to achieve a rich view of the student writers.

We plan to do more extensive Blind Rating, to even better check our reliability. We will review our approach to first grade writing in the areas of revision and editing, since revision was not a goal of the first grade teachers in our study and since editing requires quite a different approach in the early grades than later in the student's writing career.

We will also attempt to get more data on revision for high school and middle school, since we were unable to examine that area effectively in our study this year. And, we will take the necessary steps to get reading on board in our study. Much of the groundwork has already been laid for this, such as the development of a checklist for what should go into the Reading Portfolio.

Final Conclusions

Other schools who may want to base their thinking about portfolios on our work are, of course, invited to do so. It may be easier for some of these schools, if their values are similar to ours in relation to writing, to start with our terms and definitions and rating sheets.

However, over time a school or school district will want to develop its own concept of what is most important in student writing at different levels. We will also be glad to assist schools as they work toward the development of their own terms and definitions.

We feel there are six key benefits of portfolio assessment of student writing:

1. It allows professional educators to assess and evaluate according to agreed upon goals for students.
2. It allows us to do our assessing of students by looking at real work over time.
3. Interpretation of results allows schools to consider effectiveness of particular methods in their program.
4. It creates a highly beneficial teacher collaboration, including fruitful dialogue about the relationship between assessment andcurriculum goals.
5. It lays a solid foundation of information from which educators can begin to speak more forcefully and more credibly about what they and their students are achieving in the schools.
6. Finally, it increases student-writers' awareness of their growth over time, thereby increasing their self-esteem. This awareness is achieved both because the teacher more fully notices student growth and because the students themselves review and reflect upon their development.

After only the first year of our study, we are very much reassured that portfolio assessment is not only an excellent idea but one that it is quite possible to effectively implement fairly broadly, even in the busy and difficult lives of the schools. Let us hope this good news travels fast.

* The full Assessment Report is available upon request. Included in the full report are: checklists and guidelines for what should be gathered in writing and reading portfolios, student results on "growth charts" for each grade level, samples of student writing studied, and fuller clarification and analysis.

Level 1st (1-23)	AMOUNT OF CHANGE					
# Of Students Who:	FLUENCY	DEVELOPMENT	ORGANIZATION	STYLE/VOICE	REVISION	EDITING
Improved by 1 1-2				5		1
2-3	10	4	3	3		
3-4	1		2			2
4-5						
Improved by 2 1-3		7	5	6		13
2-4	4	3	4	1		1
3-5	1			1		
Improved by 3 1-4	1	5	8	4		3
2-5	3	1				
Improved by 4 1-5	1	1				1
Stayed Same 1-1						
2-2				1		
3-3				1		1
4-4	1	1				
5-5	1	1	1	1		
Worse 3-2						1

Level 1st (24-51)	AMOUNT OF CHANGE					
# Of Students Who:	FLUENCY	DEVELOPMENT	ORGANIZATION	STYLE/VOICE	REVISION	EDITING
Improved by 1 1-2	6	4	4	7		10
2-3	4	3		2		
3-4			1			
4-5						
Improved by 2 1-3	6	11	17	14		2
2-4	4	2	1			
3-5						
Improved by 3 1-4	2	4	2	2		
2-5	2					
Improved by 4 1-5	3					
Stayed Same 1-1		3	3	3		15
2-2	1	1				1
3-3						
4-4						
5-5						

Appendix C

Reading Portfolios: An Effective Means for Assessing Growth of Student Readers

Appendix C presents the key "steps" and "tools" necessary to gather and rate reading portfolios similarly to the way we rated writing portfolios. In this case we provide the "Terms and Definitions" first, to orient you to the criteria we chose to apply in reading, before we offer the "Teacher Checklist" for gathering reading materials and the subsequent "Reading Portfolio Assessment Rating Form" and "Grade level rating values" for a primary level and high school level. Examples of Reading Logs and Reading Response Journal topics are also provided.

Overview

From the start of our three-year project we were interested in reading as well as writing. We realize that writing and reading are two sides of the same literacy coin, that both are needed for success in working with language. However, we also made a conscious decision to allow our work to lead with a central focus on writing. We wanted to accomplish first, as we have said in the main body of this book, a systematic and rich treatment of writing. Only after we felt secure that we were, in fact, achieving our goals with writing, in the third year of the original project, did we turn our attention more intensely to reading.

We were able at that point to pilot the enclosed materials developed for examining reading development. We implemented our criteria and rating system with a core group of students at two grade levels. Though our data is rudimentary, it does begin to support our belief that portfolios can be a reliable method for gathering information and for monitoring student reading development effectively.

At the grade levels studied, our rating system worked well to document the amount of progress each student made in the various aspects of reading. Because in our workshop classrooms students were provided multiple opportunities to read a variety of texts and to respond to that reading in meaningful ways, we were able to collect a rich amount of student work from which an assessment could take place.

Reading logs provided daily evidence of the amount and of the variety of materials read. Response journals, anecdotal records, assigned tests, and individual projects documented how well students understood their reading and how well they were able to apply their knowledge to other situations. Affective surveys and journal entries spoke to a student's feeling about the reading process and allowed for a rating in the area of "Pleasure."

So, even though we got to reading late in the game of our research, with these pilot groups we were able to develop a solid framework with its own rating scale criteria. This allowed us to assess reading for a primary level classroom and an advanced high school classroom. Because we required ourselves to develop the "system" for assessing reading, we believe those materials and this appendix can provide guidance for reading similar to the guidance for writing provided by the rest of the book.

Here again, we think our work can provide a map to others about how to do this. You may wish to assess other features of reading than we chose to examine. And, you may end up with quite different rating sheets and so forth. You will need to assess what you value in your teaching and your curriculum in relation to reading. Though our two sets of rating values are grade level specific, we feel these materials can be easily reworked to meet the needs of many. In addition to the materials included here, commonly used diagnostic methods such as running records would be helpful at elementary grades.

What follows, then, as referred to in our appendix introduction are "Reading Portfolio" materials and guidelines. We'll walk you through them. The first several pages we call "Section 1." Those are the materials needed for actual rating of Reading Portfolios, namely: "Terms and Definitions," a "Teacher Checklist" for gathering student reading portfolios, a "Reading Portfolio Rating Form," and a "Rating Values Sheet" for second grade and for eleventh grade.

The second selection of materials, "Section 2," includes: "Reading Log Forms" for kindergarten and other primary level grades, "Reading Log Forms" for middle school and high school use, "Reading Response Journal" prompts for elementary levels that help students respond in a variety of ways to different genres, and prompts that help students respond by noting evidence in their reading of Art Costa's different types of "Intelligent Behaviors." Elementary teachers in our research study attempted to teach students differences among different genres and also desired to help students learn to recognize and apply the different intelligent behaviors in their reading and in their lives.

Section 1: Terms and Rating Forms

Reading: Terms and Definitions

Amount

Rationale: An important aspect of a student's reading is the amount of reading. To improve fluency in reading, to help achieve variety in reading, and to acquire the skillful approaches to reading needed to function well in society a high amount of reading needs to occur during schooling.

Definition and Guidelines:

- How much reading has the student completed altogether?
- How much time beyond reading class has the teacher provided for additional reading?
- How much additional reading has the student done?

Reading logs are helpful in determining the amount of material each student has read. The record provides valuable information regarding dates, titles, genres, and the number of pages read each day. (See example Reading Logs in Section B on pages 133–135).

Comprehension

Rationale: It is important for students to understand what they have read in order for the reading experience to be successful. It is also important for students to be developing strategies and approaches that will help them effectively read increasingly complicated and demanding texts over time.

There are at least two levels of comprehension that should be examined, "surface comprehension" that relates to the literal information and basic purpose of the text, and "deep comprehension" that involves being able to analyze subtler meanings of the text and also apply those meanings and values to life situations (this last area examined under "Connecting to Life and Values").

Definition and Guidelines:

- How well has the student understood what has been read?
- Does the student understand the literal level of meanings in the text—plot or chronology, emphasized character roles, facts and basic information?
- Does the student understand the deeper meanings and issues of the text—cause-effect relationships, comparison and contrast purposes, motivation for character decisions, underlying values or lessons being expressed, implications of central insights of the text?

Pleasure

Rationale: It is important for students to develop pleasure in reading and to deepen that pleasure over time so as to become avid readers. This development of pleasure in reading is essential because only if reading is pleasurable will students read the amount and variety of books that can cause them to become highly skillful readers, and only if pleasure occurs and grows will students achieve the goal of becoming lifelong readers. Only through lifelong reading can our students grow into the type of adults who achieve fuller and more wide-awake lives through reading, and only through lifelong reading can our students become the kind of people who can make wise decisions as citizens in a democracy.

Definition and Guidelines:

- How much does the student enjoy, or appear to enjoy, the reading experience? This question can be asked of different experiences, such as: being read aloud to, choral reading, reading in pairs, silent reading, reading aloud, reading at home.
- Most directly, is there a clear note, or tangible record, of the student's pleasure in reading?
- Do there seem to be periodic changes in the student's level of pleasure in reading?
- Do affective surveys show a certain level of pleasure in reading?
- Types of expression that can be observed to note levels of pleasure:
  - —Does student express hostility and distaste? Boredom and disinterest? Excitement and fun in regard to the reading experience?
  - —Does the student demonstrate involvement in "reading extension" activities or announce a desire to "share" with the class or others about reading?

Connecting to Life and Values

Rationale: It is important for a student to see the worth of reading beyond comprehension and pleasure. It is also important for the student over time to come to exhibit reading habits and abilities that a "skilled adult" exhibits. A key ingredient of these habits or abilities is the applying of lessons or values from a book to life experiences. To benefit fully from reading, students must realize there is potential connection between text and life.

Definition and Guidelines:

- How much, how well, and how often does the student connect the reading to her/his own life experience and to questions of value raised by conflict, disagreement or struggles in the texts?

The expressions of this connection will probably be in levels, as with pleasure. Three levels might include: no expression or implying of a sense of connections between text and life; some general sense of connection, or a clear sense of connection on a central point; and, then, at a more skillful level, connections would be represented by higher level thinking applied to the text and related life situations (hypothesis development, analysis, evaluation, application, establishing priorities, extending to other situations beyond just the life of the reader, addressing values changes and personal philosophy development).

There may be no way of knowing "how often" a student has done these steps. However, a rich collection of student responses both written and oral are helpful in monitoring the depth to which a student connects to her/his reading.

Reading Portfolio Assessment Form

(effectiveness) (amount used)	ineffective very little 1	low little 2	moderate moderate 3	good frequent 4	excellent exceptional 5
Amount					
Comprehension					
Pleasure (expressed directly or indirectly)					
Connecting to Life and Values					
Variety					
Academic Reading (usual school tasks)					

Teacher Checklist for Gathering Student Reading Portfolios

1. On a selected class day, probably the day after the Writing Portfolios are gathered, the teacher asks students to help generate, or complete, the Reading Portfolios.
2. The students bring up to date and add any necessary clarifications to their "report" on how much and what pieces they have read.
3. Then, the students also place in the portfolio any materials related to their reading (labeling the materials and double-checking to see that all materials are dated). The additional materials might include:
   - writings done in response to readings (informal and formal writing, reading logs, etc.)
   - reading extension activities completed—if these cannot all be placed in the portfolio, the students can describe the activities and any products as carefully as possible.
   - and, *any* other reports or teacher assessments of student's reading skills or reading development (anecdotal records, conferencing report summaries, test results, etc.)
4. Then, the students write a letter of introduction to the Reading Portfolio, which includes:
   - what readings the student enjoyed most, and why
   - what readings the student enjoyed least, and why
   - what the student likes best about reading
   - what the student likes least about reading
   - how the student feels about her/his reading development, and future plans for reading
5. Then, the teacher writes a letter of introduction to the Reading Portfolios, which includes, in brief:
   - a clarification of the context or assignments the student reading came from
   - her/his goals for the students as readers (early, middle, and late in the year)
   - her/his explanations of other materials in the portfolio, or of extension activities not in the portfolio
   - the teacher's assessment of one high and low portfolio using the Assessment Rating Form
6. All materials are gathered in a box and clearly labeled by School, Teacher, and Class.
7. Affective Surveys from early and late in the year are labeled and added to the box.

Second Grade Reading Values

Name ______________________ Fall ❑ Winter ❑ Spring ❑

Amount

1. 0–9 books
2. 10–14 books
3. 15–18 books
4. 17–18 books
5. 19+ books

Comprehension

1. No evidence of understanding
2. A few facts about the story
3. Facts depicting character, setting, and plot
4. Interpretation of story describing characters, answering "why" questions; a depth to response
5. Facts and an in-depth analysis of story

Pleasure

1. Expresses a dislike for reading
2. Shows a general satisfaction with reading
3. Enthusiastic expression of pleasure

Connection to Life

1. No evidence of connecting story themes/ideas to personal life
2. Evidence is brief and vague
3. Clear evidence of connecting reading to life; more fully stated
4. Relates themes/ideas more fully to personal life; provides examples
5. Detailed connections to personal life; provides examples; possible synthesis

Eleventh Grade Reading Values

Name ______________________ Fall ❑ Winter ❑ Spring ❑

Amount

1. Less than the required amount of reading
2. Meets the required amount of reading
3. More than the required amount

Comprehension

1. Unable to discuss any specific points in the story; minimal understanding of some parts
2. Able to understand basic story line; possible minor conversions; little or no detail
3. Clear and accurate knowledge of story with reference to detail
4. Further elaboration of specific events along with some understanding of significance of the events
5. Thorough and complete understanding; able to analyze details

Pleasure

1. Expression of dislike
2. Does not like the reading generally, but finds some aspect of it to like
3. General satisfaction
4. General satisfaction with enthusiasm for aspects of the reading
5. Enthusiastic expression of pleasure or an implicit sense of enjoyment

Connecting to Life and Values

1. No connections evident or simple one sentence connections
2. Brief and vague connections, somewhat elaborated
3. Attempting detailed and appropriate connections, multiple examples, though still somewhat superficial
4. Solid connections, sees significance, but not yet fully elaborated.
5. Perceives significance and implications, draws conclusions that reflect on life; complete in every way

Section 2: Reading Logs and Journals

Draw a picture to show about your book.

Name ______________________ **Date** ____________

Title ______________________________________

Tell about your book here.

Reading Log

Date	Time (spent reading)	Title	Pages	Recommended by other student (O) Assigned by Teacher (T) Your choice (C)

Reading Responses	Response Focus: What did this book make you think about? Have you had experiences like any of the characters in the book?

Reading Log				
Date	**Title of Book**	**Author**	**Genre**	**Number of Pages**

Characteristics of Intelligent Behaviors

Source: Art Costa

1. Persistence/Persevering
2. Decreasing Impulsivity
3. Listening to Others with Understanding and Empathy
4. Flexibility in Thinking
5. Metacognition: Aware of Our Own Thinking
6. Checking for Accuracy and Precision
7. Questioning and Problem Posing
8. Drawing on Past Knowledge and Applying it
9. Precision of Language and Thought
10. Using all the Senses
11. Ingenuity, Originality, Insightfulness: Creativity
12. Wonderment, Inquisitiveness, Curiosity, and the Enjoyment of Problem Solving

What Do Intelligent People Do?

(Adapted from the Characteristics of Intelligent Behavior by Art Costa)

1. DOUBLE CHECK to make sure the work they do is their best.
2. LISTEN to the thoughts and ideas of others with an open mind
3. USE GOOD LANGUAGE when speaking and writing. Language is clear and the message is to the point.
4. CREATE new things and new ideas. Take risks and learn from mistakes. Keep trying.
5. PERSIST and never give up. Keep working at a problem, always looking for a new solution.
6. PLAN before important work is started. Work hard and adjust the plan
7. CARE about people and the environment. Demonstrate kindness and respect
8. WONDER about life around them. Ask good questions and think about things in new ways.

Reading Response Choices

Fiction

1. List six questions you would ask a character in your book. Remember to use capital letters and question marks.
2. Pretend that the book you are reading is being made into a movie. Who would you cast in the leading roles? Give the name of the book characters and the name of the actors/actresses you have chosen to star in the movie. Give a short explanation of your decisions.
3. Write a book review of the selection you are reading. Tell readers why they should or should not read the book. Be specific in your explanations.
4. Write a letter to the author of the book you are reading. Explain your feelings about the book. Tell the author your favorite part of the story. Ask the author some questions.
5. Write a letter to a teacher about the book you are reading. Tell the teacher what the book is about. Tell the teacher what you liked or disliked about the book. Tell about the characters.
6. Fill out a copy of the basic book report form.
7. Predict the future of the main character in your story. What do you think the character will be doing twenty years from now? What kind of job will the character have? Where will they be living? What changes have occurred in their life?
8. List five words to describe the main character in your book. Tell why you chose each word.
9. Compare a character in the book you are reading to a character in another book you have read. How are the two alike or different? Give specific explanations.
10. What intelligent behaviors do the characters in the book demonstrate? Name the behaviors and explain how they are evidenced in the story.
11. What is the conflict in the story you are reading? How can the problem be solved intelligently? Give details in your explanation.
12. If the main character in the story you are reading had $100, what would they spend it on? Why?
13. What other titles might the author have named the book you are reading? Generate five new titles.
14. Does the theme of the book you are reading connect to your life? How are your experiences similar? What has the reading taught you? Do you think differently as a result of the book?
15. Find out some information about the author of the book you are reading. List some interesting facts about their life. Are any of these facts reflected in the author's writing?
16. Would the main character in the book you are reading make a good friend? Tell why or why not.

17. Have you read other books by the same author? How are the books similar? What do you see as the author's strengths?
18. Does the story you are reading take place in an interesting location? Find out more information about the setting of the story.

Non-Fiction: People

1. Tell about the person you are reading about. Provide facts about where the person was born and how they grew up. What is this person famous for? What do you find most interesting about them? How are you like this person?
2. How does the person you are reading about compare to another person you have read about? Give examples of how these people are alike. Have they "touched the world" in similar ways?
3. What intelligent behaviors does the person you are reading about demonstrate? Be specific.
4. What lessons have you learned from the person you are reading about? What has their experience taught you? What can we all learn from this person?

Index

About the Authors

Richard Koch is a Professor of English at Adrian College, Michigan, where he teaches writing and the teaching of writing. He has been Director of the Southeast Michigan Writing Project and of the network of Michigan Writing Projects. He also was a Demonstration Project Facilitator for the Michigan English Language Arts Standards Project. He has published articles about writing theory, responding to writing, and the use of standards.

Jean Schwartz-Petterson is a teacher for Adrian Public Schools, Michigan. She has taught elementary grades K–4. Currently she serves as a Reading Recovery teacher, and provides support to elementary staff with writing instruction. She has provided professional development in early literacy and in the teaching of reading and writing workshop.

Contributing Authors

Kathryn Bell is Curriculum Director for Hillsdale Community Schools, where she also teaches high school English and Journalism.

Liz Webb teaches high school English for Hillsdale Community Schools.

Aaron Schippert is a fifth-grade teacher for Hillsdale Community Schools.

Alan Willoughby is a third/fourth grade teacher for Hillsdale Community Schools.

All four of the contributing writers have written for publication previously and have also taught reading and writing workshop methods to area teachers.